THE DEVELOPED WORLD

Spencer Thomas

THE DEVELOPED WORLD

★

BELL & HYMAN
LONDON

First published in 1980 by
BELL & HYMAN LIMITED
Denmark House
37–39 Queen Elizabeth Street
London SE1 2QB

© Spencer Thomas 1980

ISBN 0 7135 1095 1

Typeset by Santype International Ltd, Salisbury, Wilts.
and printed in Great Britain by
Butler & Tanner Ltd, Frome and London.

Acknowledgments

Thanks are due to the following for permission to reproduce photographs and illustrations:

Alcan (UK), 7.14; Atomic Energy Authority, 6.10; Bay Area Rapid Transit Authority, 3.14; B. J. Berry, 5.9; B. Birch, 9.5; J. Bird, 6.17; B. P., 6.12, 6.19; British Rail, 5.2, 5.3, 5.4; British Steel Corporation, 7.2, 7.3; Gerd Busse, Gunther Hartung and Helmut Brauer, 9.9; M. Carr, 4.16; Copenhagen Municipal Authority, 4.17, 4.20; P. Corrigan, 3.20b; Countryside Commission, 8.6, 8.11; Claus Dahm, 7.4; R. Dalton, 9.14; Department of the Environment, 3.22, 8.1, 8.2; Department of Industry, 7.11; Der Stern, 5.13; EEC Documentation Centre, 6.7, 9.13; B. Ehrenfeuchter, 7.5; French Government Office, 6.21; Glenrothes Development Corporation, 7.13; P. Haggett, 5.5; M. Eliot Hurst, 7.6, 7.7, 7.8; I.R.Y.D.A., Madrid, 9.4; P. N. Jones, 4.10; D. E. Keeble, 7.12; Mrs N. Kingbourne, 8.3; R. Lawton, 3.13; Lea Valley Regional Authority, 8.12; R. Lester, 6.16; London Transport, 3.12; Milk Marketing Board, 9.15; Milton Keynes Development Corporation, 3.19a; Netherlands Geographical Institute, Utrecht, 4.22, 4.23, 9.7; Nottingham City, 5.12; A. Reed, 8.4; R. Riley, 6.9; Runcorn Development Corporation, 3.19b; Scientific American, 1.1; South West Planning Council, 8.5; Tesco, 4.3; J. Tuppen, 3.20a; Volkswagen A.C., Wolfsburg, 3.21; K. Warren, 6.15, 7.16; Water Resources Board, 5.11; G. P. Wibberley, 8.10.

Thanks are also due to Andrew for help with photographs, to Audrey for proof reading, to Sarah for help with the index, and to Andrew Reed for stimulating advice and criticism during the book's gestation period.

Contents

Preface

During the past decade Geography has undergone a conceptual revolution with an intensive reappraisal of its nature, spirit and purpose. There has been a parallel ferment of curriculum evaluation and innovation, and changes in examination syllabus and methods of assessment. This has given the teacher a greater degree of freedom and responsibility. At the same time there has been a fundamental reorganisation of schools structures and a reconsideration of the nature and purposes of teaching and learning. Consequently it is not surprising that at present there is a wide range of practice in school geography from an uncritical adoption of the new approaches to a persistence with courses which are conceptually outmoded. *The Developed World*, along with its companion volume *The Developing World*, is written in this context and is designed to equip teachers to meet the challenges brought about by the changes in curricula, schools and society.

There has been an identifiable movement away from the traditional syllabus with its emphasis on regional content examined predominantly through the recall of facts, to the thematic syllabus built around concepts or key ideas emphasising skills and techniques. The abilities of manipulating data, reasoning, explanation, hypothesis testing, understanding and the encouragement of opinion, attitudes and judgements are now as significant as the memorisation of facts. These concomitant shifts in content and methods have placed an almost intolerable burden on teachers.

This book is the outcome of many years' attempts to wrestle with these issues. Initially the absence of text books meant the production of domestic materials, usually worksheets. Many of the pages in this book began life at the chalk face and are the product of sweat and tears and trial and error in the classroom. However it is believed that something more than a series of exercises is needed. Consequently in both volumes there is a permeating progressive framework. Here the theme is that of the Developed part of the world – its emergence, impact, extravagances, problems and consequences, and ultimately our stewardship of the earth. The main ideas are italicised in the text and it is intended that they are presented in a logical sequence.

The examples are drawn primarily from Western Europe, with which pupils are most familiar and relate most easily, and from further afield when appropriate. Experience has also shown that only rarely can identical materials be used with both GCE and CSE classes. Therefore this text is directed at the GCE band *but* the resources, photographs, maps, diagrams, extracts etc., *are* suitable for the whole age and ability range. The enterprising teacher will utilise the wealth of potential in these resources for all candidates in what is already a common examining system.

Alert readers will recognise the heritage of various curriculum development projects, particularly those sponsored by the Schools Council and guidelines published by the Inspectorate. These require the generosity of a multiplicity of individuals and firms and these are acknowledged gratefully. However the translation of enthusiasm into the stark reality of the classroom has been the lot of teachers belonging to the West Sussex curriculum development group and students at the former Bognor Regis College of Education. They will recognise their inspiration in these pages and they are already aware of my appreciation and indebtedness to them all. My family have also been the companions and contributors to much of the original material contained here and deserve an equal share of the billing.

SPENCER THOMAS

1 Development

The world is divided into different groupings. Some such as Capitalist and Communist blocs are political and ideological, others such as the European Economic Community (EEC), Latin America Free Trade Area (LAFTA), and COMECOM are economic and social, while still others such as Developed, Developing, Third and even Fourth World indicate relative positions in a world league table compiled from various data. However, even in a seemingly homogeneous country such as England or a grouping such as the EEC

1.1 Countries with food shortages and surplus, and income

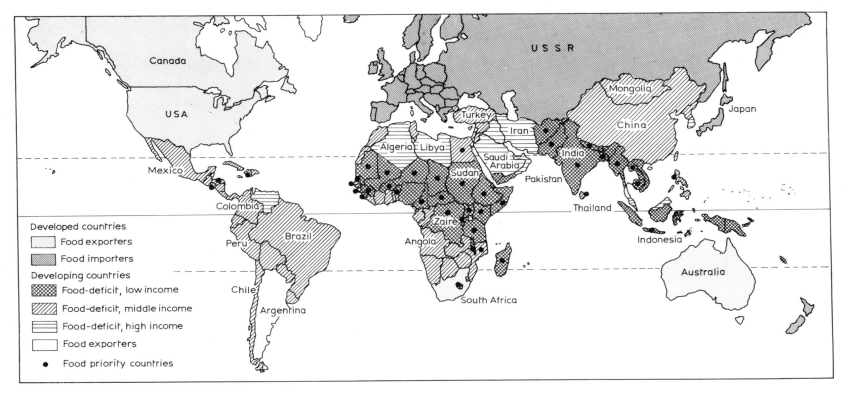

there are wide differences from region to region in income, employment, diet, health care and provision, birth and death rates and
other criteria. Therefore care has to be taken when interpreting
statistics, maps and diagrams because they generalise and often
obscure important differences from place to place. Nevertheless
they can be used as measures and illustrations of similarities, differences and trends to help make sense of a complex world. Study
the table below which gives a selection of data for certain countries
for a recent year.

Country	Income per head US$	% popn. in urban settle- ment	Annual growth rate of popn.	Output of steel (m. Tonnes)	Tele- phones per 1000	Cars per 1000
USA	7660	76	1·0	116	657	478
Sweden	7878	44	0·6	5	612	307
UK	3910	78	0·3	26	340	244
Italy	3040	66	0·7	23	229	243
Spain	2890	42	0·9	10·6	181	109
Portugal	1630	28	0·5	—	109	90
Turkey	948	18	2·5	2·3	21	6
Nigeria	440	12	3·0	—	16	4

1 Divide your page into two columns headed Developed and Developing countries. Write beneath each the countries you consider
  fall into each category. Is there a clear distinction? Or is there
  a middle grouping? Why?
2 Select one country from each column and explain how you arrived
  at your decision to place it in that category.
3 What do these statistics show about the 'stage of development'
  reached by each country?
4 The map in fig. 1.1 illustrates stages of development throughout
  the world. Describe the pattern briefly and explain the differences.
  This book attempts to describe and explain the characteristics
of the Developed World, while the parallel book does the same
for the Developing World. In addition to fig. 1.1, figs. 6.13, 7.6,
7.7, 7.8 and 7.9 provide immediate comparisons between countries
in the various categories.

2 *The Population Explosion*

It took from the beginning of mankind to 1800 for the world's population to reach 1000 million. The next 1000 million came in just over 100 years. The third 1000 million came in 30 years. The fourth 1000 million came in 15 years by 1975. By 2000 AD there will be 7000 million people in the world – over half as many again as there are now. It means that world population is growing by 100 million a year – a new USA every 2 years! (fig. 2.1)

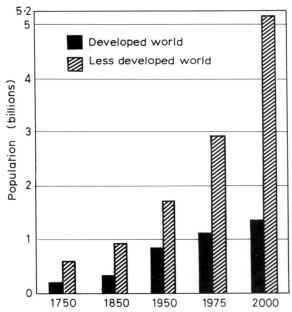

2.1 World population increase 1750 – 2000 AD

THE GROWTH OF WORLD POPULATION

Time period	Total world population in millions	Rate of population increase per 1000
The beginning of mankind about 1 million years ago to 8000 BC	8	0·08
8000 BC to AD 1	300	0·36
AD 1 to 1750	600	0·56
1750 to 1800	1000	4·4
1800 to 1850	1300	5·2
1850 to 1900	1700	5·4
1900 to 1950	2500	7·9
1950 to 1975	4000	17·1
1975 to 2000	7000	19·0

1 What has happened to world population over the last 2000 years?
2 What has happened to the rate of increase of world population over the last 2000 years?
3 Between which dates was the largest rate of increase of the world population recorded?
4 By how much did the total world population increase over the same period?
5 How can you explain the fact that total world population will increase by a larger amount over the 25 years 1975–2000 than in the period you noted in your answers to questions 3 and 4, although the rate of world population increase between 1975–2000 is predicted to be much less in this period than in the period you noted in your answers to questions 3 and 4?

Pouring a quart into a pint pot

The present rate of population increase – approaching 25 per 1000 or $2\frac{1}{2}$ per cent per annum – is almost certainly without precedent. We cannot be absolutely certain because there has never been a census of all the people in the world. The first official census in any country took place in Sweden in 1750. Britain had one in 1801. Even censuses in countries in the Developed world are subject to errors and many countries, mostly in the Developing world, have still not held an official census. If the size of the present population is imperfectly known, that of the past is even more uncertain. It must be understood therefore that numerical analysis of world population is largely a matter of conjecture.

If the conjecture and estimates are close to being correct and the present rate of increase of world population is maintained at or around $2\frac{1}{2}$ per cent per annum then world population will double every 35 years, be multiplied by 1000 every 350 years, and be multiplied by 1 million every 750 years.

The consequences of such a rate of increase are startling:

In less than 700 years there would only be 1 square foot for every person on the earth's surface.

In less than 1200 years the human population would outweigh the earth.

In less than 6000 years the mass of humanity would form a sphere expanding at the speed of light.

Prophets of doom

It is only relatively recently that people have begun to show concern for the pressures exerted by population growth of such magnitude. This awakening has led to campaigns to control this growth, and some Developed countries such as Western Germany adopted a policy known as *zero-growth*. In Britain organisations have drawn attention to the problem but population gravitated towards zero growth without the adoption of an official policy. But in Developing countries high birth rates of around 40 per 1000 per annum and lowering death rates of about 10 per 1000 per annum are resulting in average increases of 2·4 per cent–2·7 per cent per annum.

The growth of population

The populations of most countries grow because more people are born each year than die. This is called a *natural increase. Growth of population* is therefore the *excess of births over deaths* in any year. This is usually expressed as a percentage. For example, Colombia has a birth rate of 44 per annum and a death rate of 9 per annum. This means that in a given year births exceed deaths by 35 ($44 - 9 = 35$). Expressed as a percentage the 'growth rate' of Colombia is 3·5 per cent (35 per 1000 of the population).

Migration

Population growth is rarely as simple as this in reality. Very few populations grow only through natural increase. People enter and leave countries. Those entering are called *immigrants* and those leaving are known as *emigrants*. Some countries have large numbers of people coming in or going out and the difference has to be calculated with the natural increase to produce an accurate growth rate. For example at the census in 1968 France had an immigrant population of 2 600 000 which was 5·3 per cent of the entire population of the country. The immigrant population of France is increasing annually since the *net immigration*, that is, the surplus of immigrants who remain over those who leave, is between 100 000 and 180 000 a year. Western Germany has over 4 million immigrants who comprise 8 per cent of the total population.

A human tidal wave

The tide of human beings flooding to the cities is not confined to the Developing world. Cities in the Developed world generally experienced growth rates similar to those of the Developing nations today during the late eighteenth and nineteenth century when people

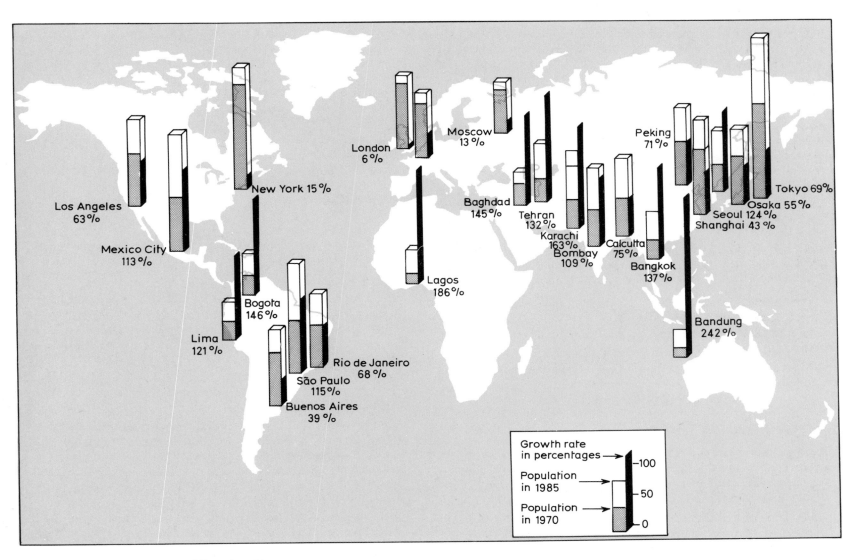

2.2 Population growth in the world's major cities

were flocking from the countryside to work in the new industries:

Manchester grew by 45 per cent between 1821 and 1831.
The population of Leeds grew by at least 50 per cent in every decade from 1811 to 1851.
Middlesborough only had 4 houses in 1801, 40 inhabitants in 1829, over 5000 inhabitants in 1841 and over 100 000 by the end of the nineteenth century.

Even so, none can parallel the phenomenal increase predicted for so many cities in the Developing world, as fig. 2.2 illustrates.

Disguise

When statistics of population growth are quoted for a country, they can convey a false impression that all the country experiences this growth. In 1960 when the world's population was 3000 million, two thirds of the people made their living on the land. By 2000 AD when there will be 7000 million people in the world the rural share will be down to one third. Therefore statistics quoted for countries conceal the most important features of the population in the world today.

Over 90 per cent of the increase in world population between 1960 and 2000 AD will be in Developing countries.
Two thirds of the increase in world population between 1960 and 2000 AD will take place in urban areas.

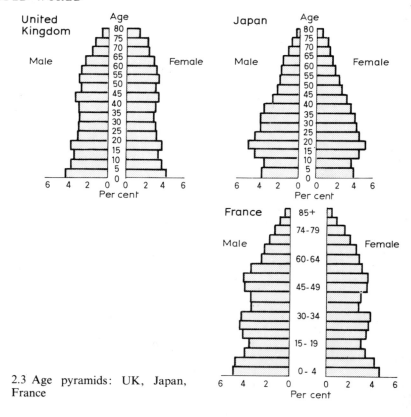

2.3 Age pyramids: UK, Japan, France

Age composition of the population

The statistics of total population and the proportions in urban and rural areas obscure further significant features about a population. Most important is the *age structure* of a population. These are usually represented as age pyramids such as the ones in fig. 2.3.

Age pyramids enable an immediate distinction to be made between the *active population* and the *non-active population*. The active population is that between the end of compulsory schooling and the beginning of compulsory retirement. In our country this definition would involve those between 16 and 65 for men and 16 and 60 for women. In the Developing countries such a distinction would be meaningless since many would not go to school and the majority would die before reaching our age of retirement, and in any event pension schemes are non-existent. The non-active population is comprised of those in school and retirement.

The shapes of the age pyramids inform us about other aspects of a population. The age pyramids of France in fig. 2.3 show the effect of two World Wars. Their main use, however, is as an indicator of the *stage of development* of a country. Countries

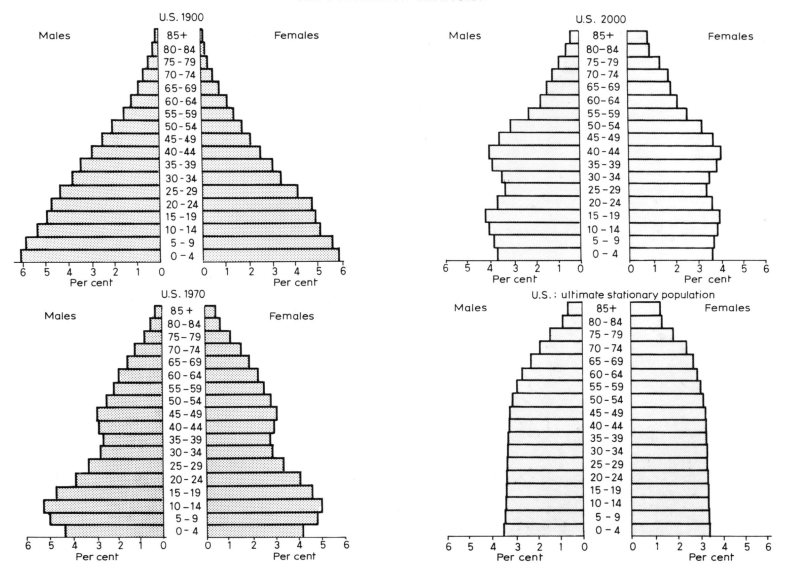

2.4 The demographic transition in the USA

which are growing fast have pyramids which are much wider at the bottom than the top. (fig. 2.4)

In contrast, the United Kingdom has an *ageing population*. Nearly one third of its population is over 60 years of age. The pyramid in fig. 2.3 shows very little change from top to bottom. There are comparatively few children because there are few women of the age when they can have children, and they are having smaller families.

Japan has a growing population which is likely to grow less quickly in the future. Fig. 2.3 shows that there was a 'baby boom' after the Second World War (1939–45) so that there is now a high proportion of the female population in the child-bearing age groups. However they are not producing sufficient children to replace themselves. This could be due to such things as the introduction or increased use of contraceptives, the legalising of abortion, a change in attitudes, a change in the economic circumstances of the country so that people cannot afford to have so many children, or a variety of other reasons. The pyramid of this type of population age structure is thin at the base with a bulge in the middle before it levels off again to assume the expected pyramidal shape at the top.

Demographic transition

The very high rates of population increase currently being experienced by many Developing nations are unlikely to be sustained in the long run, and the ultimate tendency is towards *zero growth* or *equilibrium*. The process by which this is achieved is known as the *demographic transition*. The changes in *fertility* and *mortality* which constitute the demographic transition accompany a nation's progression from a largely rural, agrarian and partly literate society to a primarily urban, industrial and literate society.

Virtually all of the countries classified as 'Developed' have passed through these stages, although the timing and the extent of these changes have differed considerably. The age structure of the population of the United Kingdom and Japan (fig. 2.3) illustrate the pyramid to be expected when the population is approaching equilib-rium. Other advanced countries such as the USA are still in the process of demographic transition as the sequence of age pyramids (fig. 2.4) shows.

The demographic transition is likely to take far less time to complete in the Developing countries than it has taken in the Developed countries. The example of the Developed countries is there as a model for the other countries and the means are available to institute the necessary changes. These means were not available to the Developed countries when they were at a similar stage of economic and demographic transition. This is especially true of birth control techniques and medicine. That is not the same as saying that the transition will be easy. The key to the reduction in the present rates of increase is education and this is far from being universal or accepted. Finally, it must not be assumed that the Developing countries will want to imitate the Developed countries. However it seems inevitable as they pass from predominantly rural agrarian societies to predominantly urban industrial societies that they will be unable to escape the same pressures and influences which will be reflected in similar responses to reproduction.

3 Settlements

Urbanisation

As late as 1900 Great Britain was the only country to be predominantly urbanised with 21 per cent of the population living in cities of over 20 000. Now all industrial nations are highly urbanised. In Great Britain cities of over 20 000 house over 80 per cent of the population, and, although just over 20 per cent of the world's population lives in urban settlements, this percentage is accelerating rapidly, particularly in the largest cities in the Developing countries (fig. 2.2).

1 Which cities had the largest populations in 1970?
2 How many of these cities were in the Developed and how many in the Developing World?
3 Which cities will have the largest populations in 1985?
4 How many of these cities will be in the Developed and how many in the Developing World?
5 Do cities in the Developed or Developing World have the largest per cent increase in population between 1970 and 1985? Why?
6 What do you think might be the consequences of such rapid rates of population growth?
7 If there is a projection or prediction for the future population of your town or district, find out the factors that were taken into consideration.

The industrial city

The Romans had mined extensively and established industries, such as glass making, in many centres. The mediaeval period had given birth to the craft guilds whose members practised a wide variety of industries, while textile manufacture steadily grew in importance. In addition, by 1640 over $1\frac{1}{2}$ million tons of coal had been mined in England and Wales. However at this time charcoal was the chief fuel, and iron working was prominent in places such as the Weald in south-east England. Restrictions on the availability of timber for charcoal and inventions which opened up the possibility of using coal as a substitute launched what has since become known as the Industrial Revolution. The impact and tremendous changes which it set in motion are summarised in fig.3.1.

1 Between the years 1701 and 1750,
  a Which was the largest city?
  b Which was the second largest?
  c Which were the other important cities? What did they have in common?
2 Which was the most populated city in 1801? Was it the same city with the largest population in 1701 and 1750?
3 How many settlements had by 1801 exceeded the population of the city with the second largest population in 1750? Which settlements were they?
4 In the eighteenth century which settlements
  a grew rapidly and
  b hardly grew?
5 Why is population growth generally faster in urban than in rural areas?

The towns and settlements which grew up during the Industrial Revolution still have the largest populations today. The conversion of Britain from a 'green and pleasant land' to the 'workshop of the world' was accomplished at great cost. Conditions such as

Population growth   **(A)**

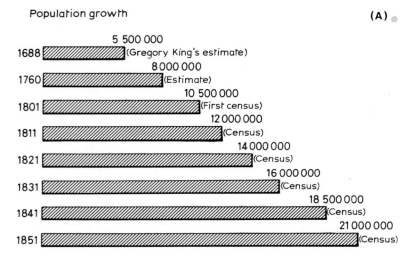

those described below were found in most urban areas during the late eighteenth and throughout the nineteenth century.

'Crowded like penned animals in small back to back houses built as cheaply as possible, sometimes with walls only a brick thick, deprived of light, overrun with vermin, lacking any recreation space, and with the all-pervading stench of human excreta, the working class of the early industrial towns lived lives of utter degradation.'

1 Many people were concerned about these wretched conditions and campaigned vigorously to remove or improve them. Imagine that you were one of these social reformers and design a poster which would shock important people into bringing about improvements.

3.1 Growth of urban settlements in Britain during Industrial Revolution

Population distribution   **(B)**
1701

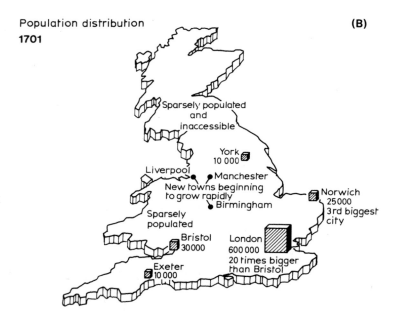

Population distribution   **(C)**
1750

London had a density
of 300 persons
per 100 acres

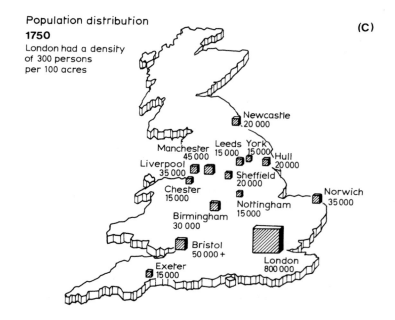

2 Address a protest meeting on the subject of social reform. Your teacher may allow the class to act as the audience and select certain members to act as particular people such as the factory owner, The Member of Parliament, the shopkeeper, the local authority health inspector, the doctor, the vicar, and some of the ordinary people living in such atrocious conditions. Make your speech and allow the audience to interrupt and question you.

3 Why did people remain in cities if conditions were so bad?

Changes in land use

Fig. 3.2 overleaf shows the changes in the land use of Britain during this century.

1 Which category of land use
  a covers the greatest percentage of land?
  b has had a continuously decreasing percentage of the land area since 1900?
  c has had a continuously increasing share of the land area since 1900?

Urban land use has increased at the expense of agricultural land, while the other major users of land have had a relatively stable percentage of the land area. At the time of the first census in 1801 there were just over 12 million people in Great Britain. In 1971 there were 55 million. It is hazardous to predict future trends, but it seems likely that to accommodate the population

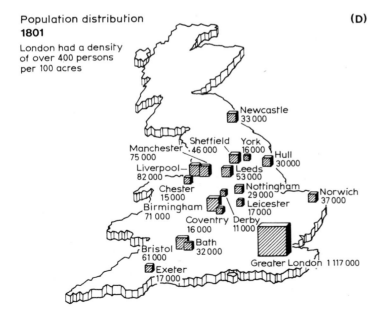

Population distribution **(D)**
1801

London had a density of over 400 persons per 100 acres

Newcastle 33 000
Sheffield 46 000
York 16 000
Manchester 75 000
Hull 30 000
Liverpool 82 000
Leeds 53 000
Chester 15 000
Nottingham 29 000
Norwich 37 000
Birmingham 71 000
Leicester 17 000
Coventry 16 000
Derby 11 000
Bristol 61 000
Bath 32 000
Exeter 17 000
Greater London 1 117 000

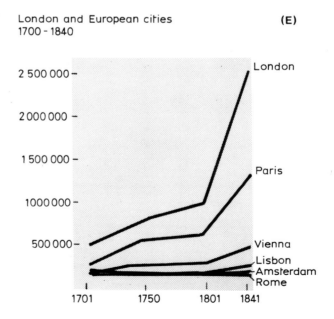

London and European cities **(E)**
1700 - 1840

London
Paris
Vienna
Lisbon
Amsterdam
Rome

2 500 000
2 000 000
1 500 000
1 000 000
500 000

1701   1750   1801   1841

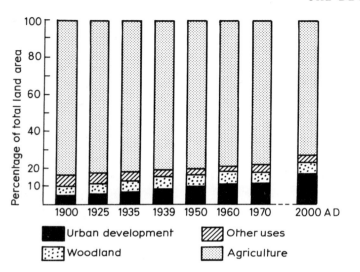

3.2 Land use changes in Britain – 20th century

3.3 Right, a traditional
village preserved. Below,
a village school now
closed

of Britain by 2000 AD it will require the building of a Wolverhampton
every four months or a Leeds every year. This will involve a
further 5 per cent of the land of Britain being lost to urban uses
bringing the total land area buried under urban settlements to
16 per cent by 2000 AD.

The disappearing village

One of the casualties of urbanisation has been the village (fig.
3.3). A survey into the English village produced the following
typical observations:

Headley, Surrey

[Village] generally felt to be in slow decline because (1) primary
school closed (2) baby clinic closed (3) bus service reduced to
one route so doctor's main surgery cannot be reached by public
transport. . . .

3.3 Right, a traditional
village preserved. Below,
a village school now
closed

Easton Royal, Wiltshire

Young people tend to leave and seek accommodation nearer places of work . . .

Uldale, Cumbria

Taking away 11 -plus children has truncated social life in the village. They often return urban-minded. . . .

Shaw Mills, North Yorks

. . . rising transport costs make commuting more costly . . .

Helsey Beauchamp, Worcs

In decline. No parson. No bus. No shop. No pub. No jobs for young people – or anybody else. . . . Grave danger school will close . . . . A good proportion of well-to-do retired people, mostly owning two cars. . . .

Nether Wallop, Hampshire

Four out of the five shops have closed in the last 8 or 9 years and the social life which comes from shopping, gossiping and keeping in touch has disappeared. [This village was referred to by the respondent as an 'estate agent's dream' in view of its beauty – and perhaps also because the pattern of development has tended to exclude low income, less 'select' families, thus probably allowing an extra premium to be added to house prices in the village.]

1 Assume you are either a long standing resident born and bred in a village or a recent newcomer who has achieved his life's ambition by buying his dream house in a rural village. Write a letter to a friend you have not seen for years, describing the village as it appears to you.

It is ironic that as the character of traditional village life is disappearing, planners respect its virtues of community feeling and mutual support and are trying to recreate it in new developments such as suburbs and new towns.

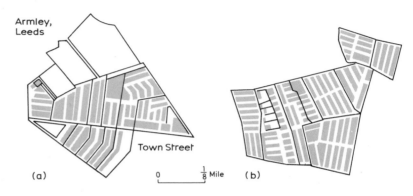

3.4 Field boundaries influence street patterns

Models of urban growth

In the early stages of the growth of towns, builders bought plots field by field. Although most of the former field patterns have been obliterated by building, it is often possible to trace their influence on the present day pattern of roads and houses as fig. 3.4 shows. Today they purchase entire farms or estates and field boundaries are removed before building begins.

It is possible to reconstruct the growth of settlements piece by piece by referring to old maps and documents such as sale catalogues, although it would take a long time. However, most show similar patterns of growth, and the main points can be summarised in these diagrams (figs. 3.5, 3.6 and 3.7).

Fig. 3.5 illustrates an *historical* explanation of how the various land uses come to be arranged in the way they are. It maintains that as the city grows outwards from the centre new houses are built on the periphery. Land is cheaper here than nearer the centre so the houses can have bigger gardens and more space. They attract the more affluent people who can afford these bigger properties.

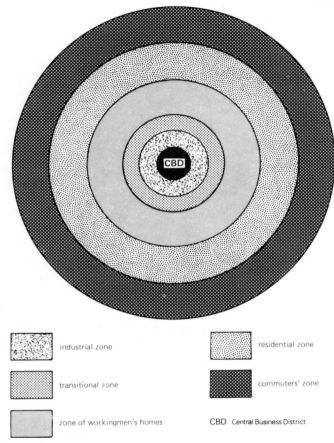

industrial zone

residential zone

transitional zone

commuters' zone

zone of workingmen's homes

CBD   Central Business District

3.5 How an urban settlement develops

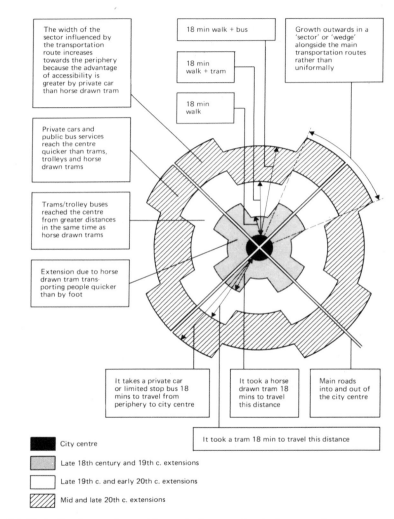

The width of the sector influenced by the transportation route increases towards the periphery because the advantage of accessibility is greater by private car than horse drawn tram

18 min walk + bus

18 min walk + tram

18 min walk

Growth outwards in a 'sector' or 'wedge' alongside the main transportation routes rather than uniformally

Private cars and public bus services reach the centre quicker than trams, trolleys and horse drawn trams

Trams/trolley buses reached the centre from greater distances in the same time as horse drawn trams

Extension due to horse drawn tram transporting people quicker than by foot

It takes a private car or limited stop bus 18 mins to travel from periphery to city centre

It took a horse drawn tram 18 mins to travel this distance

Main roads into and out of the city centre

It took a tram 18 min to travel this distance

City centre

Late 18th century and 19th c. extensions

Late 19th c. and early 20th c. extensions

Mid and late 20th c. extensions

3.6 Growth of an urban settlement along routes to town centre

The less well off who cannot afford to move out to the growing suburbs live in more cramped conditions on the more valuable and expensive land close to the centre and their places of work. This process of moving out to the edges of an ever increasing circle is known as the *centripetal* force.

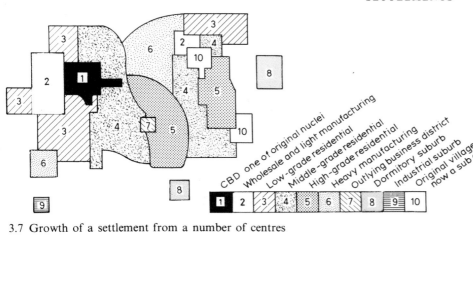

3.7 Growth of a settlement from a number of centres

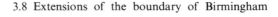

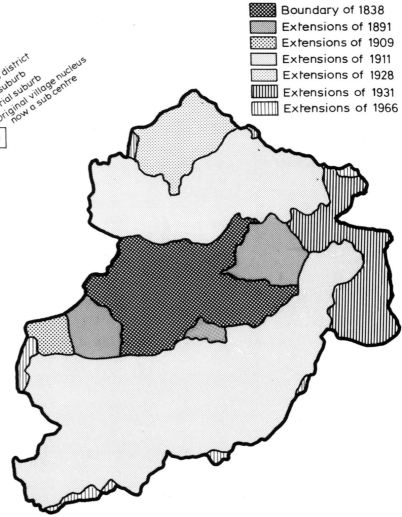

Until the early years of this century this was the predominant pattern of growth. The size of the settlement was restricted by the distance from homes to workplaces which were usually in or near the centre. However, as the means of transport improved and speed increased, as electric trams and trolley buses replaced horse drawn trams, people could live further from their work and still get there in the same time. Fig. 3.6 illustrates how this improved accessibility, especially along the main routes into and out of the town centre, altered the shape of settlements. Now, instead of growing fairly evenly all around as in the previous diagram (fig. 3.5) they extended further along the routes which became *sectors* of development. This was an *economic* explanation. However, some people observed that many settlements did not resemble either of these 'models'. Instead of developing from a single centre or *nucleus*, many places grew together or coalesced from a number of centres or *multiple nucleii*. In these settlements the land uses did not fall into a neat arrangement but became mixed up like a patchwork pattern (fig. 3.7).

A conurbation: London

Over the years and centuries, towns have grown in size and population. London has grown outwards from its centre to engulf and incorporate once separate and distinct towns and villages. In this and other parts of the country, as in West Yorkshire and South-east Lancashire, established towns with their own administrations have grown towards each other so that it is impossible to recognise where one ends and the other begins. Such a large continuously built up area is called a *conurbation*.

It is often difficult to compare the growth of one place with another over a span of years because of boundary changes (fig. 3.8). Since 1965 London has become an identifiable unit under the control of the Greater London Council, although many people would debate whether this boundary satisfactorily defined the limits of London.

Megalopolis

Occasionally conurbations grow together. Boswash, Chippitts and San-San are not the names of launderettes or Chinese restaurants! They are the names of three of the largest concentrations of population in the world. San-San is a nickname for the area stretching from San Francisco to San Diego.

1 Look at fig. 3.9 and write out the full names of Boswash and Chippitts.

2 Are there any other concentrations of population likely to merge together in the future? If so, suggest names for them.

3 Is Boswash a suitable 'sub-title' for the urban corridor along the north eastern seaboard of the USA? What are the names of the other cities involved?

4 Are there any other cities in the Megalopolitan regions of Chippitts and Boswash which are not represented in the abbreviated 'sub-titles'?

5 What is the density of population in and around the major towns?

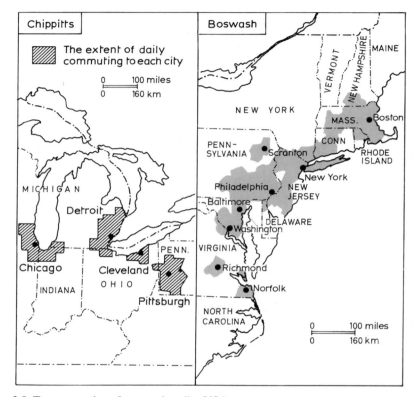

3.9 Two examples of a megalopolis, USA

Megalopolis is a term which can be applied to the built-up area which stretches from Boston to Washington (fig. 3.9). This area contains the metropolitan or city regions of New York, Boston, Philadelphia, and Washington. It only covers 2 per cent of the surface of the USA but it contains over 20 per cent (40 million) of its population. It contains the seat of government, it is the financial and business capital of the country and despite decentralisation produces 25 per cent of the country's manufactured goods. It could be termed a *primate region*.

It stretches for over 800 of the most densely populated kilometres in the world, but it is not continuously built up. The agricultural and rural recreation areas are not independent but constitute an integral part of the Megalopolitan system.

Like a spider spinning its ever-widening web and gobbling everything captured in the net, or like the tentacles of an octopus, the urban areas have reached out and engulfed everything in their paths. The city regions have become indistinguishable from each other.

Commuting

The nineteenth century immigrants flocked to the boroughs bordering the river Thames to get as close as possible to the sources of employment which were largely associated with the port (fig. 3.10). Successive waves of immigrants found it increasingly difficult to find homes in these districts and accommodation had to be found further from their workplace. As people became more affluent and mobile, they *chose* to live further from their workplace to which they travelled daily, a movement which has been called 'the flight to the suburbs'. On the other hand, the jobs became concentrated in the centre with an overwhelming preponderance in the City of London and in Westminster (fig. 3.11).

The separation of workplace, shopping centre, or entertainment focus from the places where the majority of people live has been encouraged by developments in transport. The increase in the speed of movement has propelled the outward spread of urban areas. As the suburban railway services were electrified and the underground system spread its tentacles, the transport companies attempted to lure people greater distances from their workplaces to enjoy the benefits of the countryside (and, incidentally, to use their services to get them back to work as most of the jobs stayed where they were) as the poster in fig. 3.12 illustrates.

Developments in transport emancipated people from the need to live close to their work. It changed our whole attitude towards commuting. Many people think nothing of travelling an hour or more each way each day from their homes to their work despite

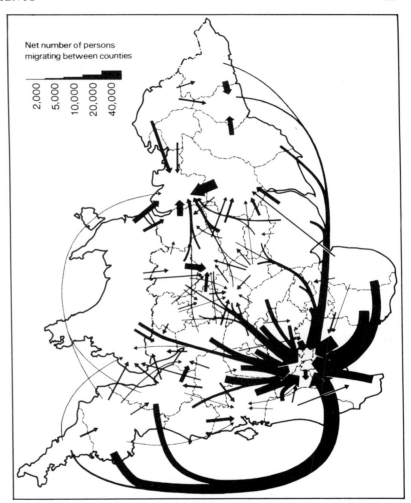

Net number of persons migrating between counties

2,000 5,000 10,000 20,000 40,000

3.10 The 19th century migration to London and other cities

increasing costs. The German word for commuting is 'Pendler'. It conjures up a vivid picture of the regular backwards and forwards daily movement like the to and fro rhythm of a pendulum. Fig. 3.13 shows the commuter zones for the major urban areas in Britain.

3.11 Job surpluses and deficiencies in the London boroughs

C   City
Hy   Hackney
Hh   Hammersmith
I.   Islington
K. & C.   Kensington & Chelsea
K.u.T.   Kingston upon Thames
L.   Lambeth
R.u.T.   Richmond upon Thames
S.   Southwark
T.H.   Tower Hamlets
W.F.   Waltham Forest
W.   Westminster

■ Enormous surplus of jobs over population

▨ Huge surplus of jobs over population

▨ Large surplus of jobs over population

▨ Relatively small surplus of jobs compared with population

All other boroughs do not have enough jobs for their own population

3.12 A 1908 advertisement designed to attract commuters

1 Look in your atlas and make a list of the places which attract commuters.
2 To which place do commuters travel the greatest distance? Why?
3 In which parts of Britain is there the greatest amount of commuting? Why?
4 In which parts of Britain is there least commuting? Why?

London has the largest commuter hinterland in Britain. Most of the 1¼ million people who commute daily into central London travel by rail or underground but a substantial proportion still use cars and buses. It is the daily ebb and flow of the human tide that creates the congestion and causes the problem with its tidal wave dimensions at the morning and evening rush hours. Despite attempts to stagger working hours and appeals to the public to avoid travelling at rush hours the problem persists. Can it be solved?

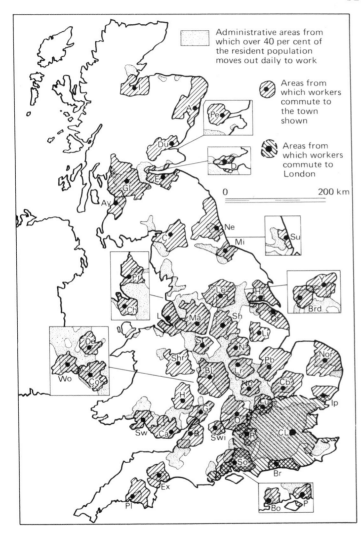

3.13 Major commuter zones in Britain

Some attempted solutions

a Ban vehicle traffic from the central areas of cities by providing large car parks on the periphery of the central areas making people walk into the centre or use public transport. In Leeds an experiment is in progress using electric battery-powered buses for this service.

b Price vehicles out of the central areas of cities. Double yellow lines, discs, and parking meters are all devices designed to deter the motorist from coming into, or spending too long in, the city centre. But more revolutionary proposals have been made, for example that all vehicles be fitted with meters like taxis. Wires sunk into the road or laser beams at the side of the road would automatically switch on the meters when the vehicles passed these points. Costs would increase progressively towards the centre where it would be prohibitively expensive.

3.14 An above-surface section of the San Francisco Bay Rapid Transit System

Some people would argue that the above prohibitive and restrictive measures are an admission of defeat. They can control but never tame the twentieth century monster. They say you should take a positive approach and make public transport systems clean, fast, reliable, and competitively priced. Then people would willingly transfer to this alternative because they get no pleasure out of driving in congested conditions. An example of such a system is operating in San Francisco.

The San Francisco Bay Rapid Transit System (fig. 3.14) is an attempt to solve traffic congestion by providing a quick, reliable, urban rail system. BART (Bay Area Rapid Transit) as it is called is fully automatic and controlled by two computers. The 80 miles an hour trains glide quietly around the 121 kilometres (75 miles) track at 90 second intervals. They run at street level, underground and on elevated stretches.

Having seen it in operation, over 70 large urban areas are considering the possibilities of introducing rapid transit systems. The first in this country is Tyneside based on Newcastle upon Tyne.

3.15 An interchange system

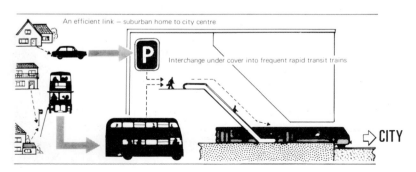

Tyneside Rapid Transit System

THE PROBLEM
Towns and cities need movement to prosper
The car adds to our ability to move
BUT
As car ownership increases
Traffic congestion gets worse
Pollution and frustration grow
AND
Public Transport deteriorates.

WHY RAPID TRANSIT?
Tyneside must have a system which moves people with
– greater speed
– greater reliability
Especially when homes and jobs are further apart than in the past
The core of the public transport system must
– be free from traffic congestion
– help to relieve the congested Tyne Bridges
– create efficient movement corridors
– let people travel directly to the heart of the built-up areas
– make the best use of the existing but expensive rail network
– be quiet and fume free
– minimise blight and property acquisition
– be capable of extension.

Rapid Transit will form the backbone of a new public transport system integrated with bus services, interchanges, and car parks (fig. 3.15).

1 How does a transport, or transit, system become 'Rapid'?
   a outline the steps by which a traditional transport system is made into a rapid transit system.
   b how does an interchange speed up the flow of people and vehicles?

2 Make a list of the possible consequences of the introduction of a rapid transit system.
3 Write a brief account indicating which measures you would introduce, and why, to meet the problem of city centre congestion.

Move the jobs

A more radical solution is to move both the jobs and the people to new locations right away from the city. A policy of moving people away from central districts has been followed by the Greater London Council and the individual borough councils to reduce residential densities. It operates in one of two main ways:

1 by entering into agreements with towns throughout England to accept Londoners as part of an 'Overspill' policy. This is not always accepted without protest by the 'host' town. Why?
2 by working with the central government to rehouse and reemploy Londoners in 'New Towns' built by Development Corporations appointed by and financed by the central Government.

The success of these and similar arrangements can be gauged by the reduction in population of London from over 9 million in 1961 to 7 million in 1981.

An alternative strategy is to transfer the jobs to where the people live at present. This is being done in Chicago and other North American cities. It has been done in Hounslow and Hillingdon where development of industry along the Great West Road and in conjunction with London Airport has produced employment opportunities.

The majority of industries which have moved from central locations to the suburbs are 'light' industries, manufacturing items which are light in weight such as electrical components. However the substantial majority of jobs in the central areas of large cities such as London are in service and office employment.

Decentralisation

In 1963 the Government set up a Location of Offices Bureau (LOB) with three tasks:

1 To ease the congestion of office employment in central London.
2 To spread offices more evenly throughout south east England.
3 To encourage offices to move to other parts of the country.

The government set a lead by *decentralising* some of its own departments. The Giro headquarters was established in Bootle and the Royal Mint was transferred to Llantrisant. In 1964/65 the government introduced strict measures to restrict the creation of office employment in London and south east England, but it was not a total prohibition. These measures had the effect of making office accommodation in central London scarcer so that it became four times more expensive.

There were a number of reasons for firms to move out:

1 Restrictions imposed by government policy.
2 'Persuasion' and assistance to move from the LOB.
3 Aggressive advertising by towns and areas anxious to attract office employment.
4 Rents of office accommodation increased steeply.
5 Savings of up to £3500 per employee on rent, rates, wages and other overheads were possible by moving out of central London.
6 The costs of commuting were increasing continuously and more people were taking up employment nearer their homes in the suburbs or dormitory towns.
7 As a result there was increasing difficulty to recruit suitable labour. Agencies opened in London offering attractive conditions and wages for 'temps' – temporary or part time work – to combat the shortage.

1 Make a list of the attractions and benefits being offered by
   a agencies anxious to retain office employees in central London.
   b centres competing for new office development and offices which were decentralising.

2 Which offers would have attracted you? Why?

In 10 years the LOB moved 1555 firms and over 110 000 jobs out of central London. They went to three main locations:

Fringe areas

At first firms were reluctant to move long distances out of central London and settled in the nearest location just outside the 'restricted' area. Just about the same time as the restrictions were imposed in 1964/65 a prime site became available in the centre of Croydon as the Trinity School of John Whitgift moved to a new site a few miles away. In addition the Croydon Corporation designated an area of 48 hectares in the town centre for office development. Rents and rates were 50 per cent lower than equivalent new office accommodation in central London. Although these have increased, Croydon still has a cost advantage as well as other facilities such

3.16 Office development in Croydon

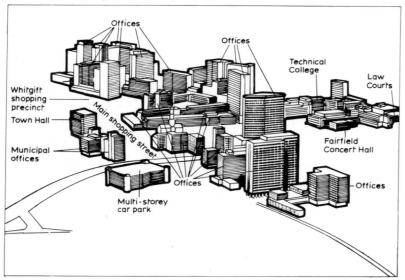

as less congestion and the proximity of a large regional shopping centre. Fig. 3.16 tells its own tale. In Croydon over 50 tower blocks house over 30 000 employees, thus making a significant contribution to reducing commuting and providing employment near the homes of those who moved to the suburbs.

Expanded towns

Under the provisions of the New Towns Act 1952, the London County Council (now the Greater London Council), the Hampshire County Council and the Basingstoke Borough Council agreed in 1961 to expand the population of Basingstoke. Costs are shared between the three authorities, which makes it distinct from 'New Towns' which are entirely government sponsored and financed. The location of New and Expanded towns is shown in fig. 3.17.

Basingstoke is the largest 'Expanded Town' scheme in the country: the Greater London Council through its Industrial Centre and in conjunction with the LOB finds firms in London willing to move to Basingstoke with most of their employees who are allocated homes for rent or purchase. The scheme has attracted well known firms and organisations. The Automobile Association (AA) which alone employs over 2000, the Civil Service Commission, Thomas de la Rue who print banknotes, the Department of Health and Social Security, Berk Ltd. (Chemicals), British Oil and Cake Mills Ltd. all have their headquarters in the town. The 22 hectare site on the northern slopes of the Loddon valley has ample space remaining to accommodate the proposed doubling of office employment by 1986. Its position adjacent to the M3 motorway and 22 kms. from an access point on the M4 motorway and 75 kms. from central London has played an important part in its success in attracting such notable tenants.

New Towns

New Towns are not such a new idea. All towns were once new and many, notably in the mediaeval period, were planned. Even among towns born during the industrial revolution of the eighteenth and nineteenth centuries there were a few model settlements.

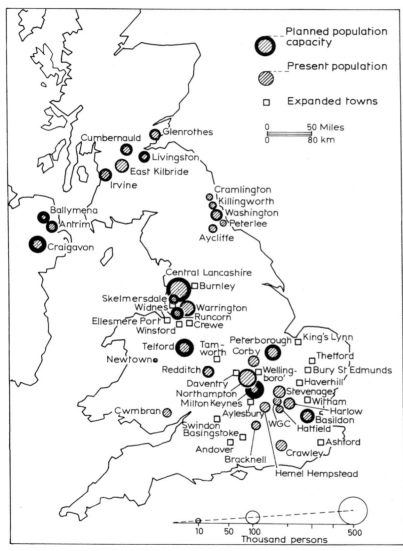

3.17 New and expanded towns in Britain: present and planned population targets

1 Robert Owen took over a cotton mill and industrial village at New Lanark in Scotland in 1799 and devoted his energies to the provision of social services.

2 Titus Salt established a factory employing 3000 about 4 miles from Bradford in the 1850s and constructed a town to house the workers. Saltaire still retains its separate visual identity.

3 Towards the end of the nineteenth century W. H. Lever built the model town of Port Sunlight in Cheshire to accommodate the workers of his soap factory.

4 George Cadbury built Bournville to house the workers of his chocolate factory around the same time.

Such planned towns were not confined to Britain, and Pullman City (1867) near Chicago, and Krupp's workers' settlement in Germany are two examples of similar enterprise elsewhere.

One of the most important pioneers of new towns was Ebenezer Howard. In 1899 he founded the society now known as the Town and Country Planning Association. It promoted the first 'Garden City' of Letchworth in 1903. This was a completely private venture. There were no immediate imitators, as Howard and his supporters had hoped, but the principles demonstrated at Letchworth, such as houses with gardens, varied layout, and landscaping were copied throughout the world. In the inter–war period the only new town constructed was Welwyn Garden City, which was also founded by Ebenezer Howard.

The success of Howard's plans led to the establishment of the Royal Commission on the Distribution of Industrial Population in 1937. A Royal Commission is a body of people which is appointed by the government with royal approval. The people involved are specially qualified to enquire into a particular problem. The Commission produced its report (the Barlow Report) in 1940. It gave a qualified welcome to the new town idea, and led to the creation of the Ministry of Town and Country Planning in 1943. This was followed by:

1 Patrick Abercrombie's Greater London Plan of 1944 which proposed a constraining *Green Belt* around London, and the overspill

of population and employment to 10 new towns 20–30 miles from London beyond the Green Belt.

2 Mathew Clyde's Regional Plan which proposed 3 new towns for Glasgow.

In 1946 the government set up a committee to investigate these proposals and within months the New Towns Act (1946) was passed and proposals introduced for the first New Towns.

Three generations of New Towns

The government appoints a Development Corporation which creates, plans and manages the New Town. Under the New Towns Act 1959, the Development Corporations are dissolved when the master plan is completed and replaced by the Commission for New Towns, which then manages the New Town. This stage had been reached by Crawley and Hemel Hempstead in 1962, and Welwyn Garden City and Hatfield in 1966. Basildon, Bracknell, Corby, Stevenage, Harlow, Redditch, Runcorn and Skelmersdale are all in the process of transferring responsibilities from the Development Corporations (fig. 3.17).

Look at the summary below and fig. 3.17 and answer the following questions:

	The earliest New Towns Mark I	The Second Generation Mark II	The Third Tier Mark III
target population	70–80 000	70 000	250 000
objectives	To accept overspill population and employment from a city or conurbation	To accept overspill population and employment from a city or conurbation	An emergent city region – a growth point
size	Neighbourhood units of 5–10 000 people	Residential districts with no separation into neighbourhood units	Townships of 30 000 people
organisation	Shops grouped into neighbourhood parades near public house, church and community centre	Corner shops and community facilities located along pedestrian ways and not grouped together	Hierarchy of provision. Town centre a regional centre, a comprehensive school and shopping facilities serve a district, smaller housing clusters focus on the primary school and limited local facilities
communication	Low provision of car parking and garages. Increase in car ownership not seen at time of building. Vehicles and pedestrians only segregated in town centre	Designed for a high level of car ownership and large number of car parking spaces. System of pedestrian-only ways with footbridge and underpasses	Hierarchy of roads – primary (urban motorways), secondary (spines and feeders), and distributors (local roads). Residential areas planned on principle of separating vehicles and pedestrians
industry	Concentrated in one industrial estate – mainly light industries	Dispersed over a number of industrial estates – mainly light industries	Industrial areas spread throughout the city – includes heavy as well as light industries
leisure	A green belt to surround the town and green wedges to penetrate to the town centre	Central and peripheral parks	Country parks of much larger dimensions offering greater variety of activities

1 How many New Towns have been designated?

2 How many Expanded Towns are there?

3 In what year was the first New Town designated?

4 Copy the map in fig. 3.17 into your notebook and write alongside each New Town the year it was designated.

5 Around which large city were the majority of Mark I New Towns located?

6 In which two other areas were there more than one Mark I New Town?

7 Where were the remaining Mark I New Towns located?

8 Near which large cities or conurbations are the Mark II New Towns situated?

9 Which conurbation is each of the Mark III New Towns planned to relieve?

10 Why are New Towns generally closer than are Expanded Towns to the large city which provides most of the labour and industries?

11 Which conurbation does not have a New or Expanded Town to relieve it? Can you suggest reasons why this is so?

12 Is it possible to make any generalisations about the target populations for Mark I, Mark II, and Mark III New Towns? Give explanations for any differences or similarities you find.

13 What does the construction of three successive types of New Towns suggest about their success or failure?

The Mark I New Towns were conceived within the recommendations of the Lord Reith Committee on New Towns constituted in 1946. They were to be 'self-contained and balanced communities for work and living'.

An example to the world

Although most of the British New Towns were conceived to relieve pressure on major cities, they are not all in this category.

1 Corby was built to accommodate people moving into the area to mine iron ore and work in a new steelworks.

2 Glenrothes in Fife was originally designed to house miners working in a new coal pit, but as coal production declined relative to other sources of energy an entirely different industrial structure emerged.

3 One of the objectives in building Killingworth in Northumberland was to carry out a scheme of land reclamation and transform a derelict coal mining area and an area of obsolete industrial activity into a desirable place to live.

4 Cwmbran in Gwent was seen as a dormitory settlement housing additional labour needed to man strategic industries, many of which had been located there just prior to and during World War II and where there was a labour shortage in the 1950s.

In addition to the overall concepts of New Town living, such as neighbourhood units (fig. 3.18) and vehicle and pedestrian segregation, other innovatory ideas have been pioneered (fig. 3.19).

The British prototype has now been imitated in many countries both in the Developed and the Developing World. The Western European examples make interesting comparisons.

3.18 A neighbourhood centre

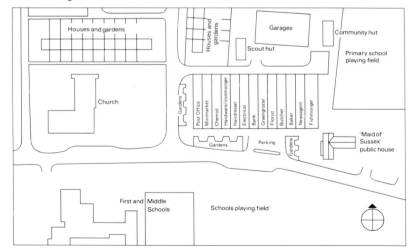

3.19(a) Bicycle way. Milton Keynes

3.19(b) Shopping city, Runcorn

New Towns in France

Until the 1960s the French answer to the increasing populations in their major cities was to construct 'Grand Ensembles' which were complexes of high rise tower blocks or 'streets in the sky' such as Le Mirail and Colomiers in Toulouse. They provoked so many problems that the wisdom of this solution was questioned. The present strategy involves the construction of New Towns, but not as in Britain designed to drain the cities of their active manpower and lively children and placed at distances which discourage the retention of links with the parent city, but designed to reinforce the position of the city. They are located close enough for the inhabitants to commute if necessary to work in the city centre, and also to encourage them to make use of the cultural and leisure facilities which only a large city can provide and so avoid expensive duplication in a New Town. As fig. 3.20 shows they are also

3.20(a) New developments around Paris

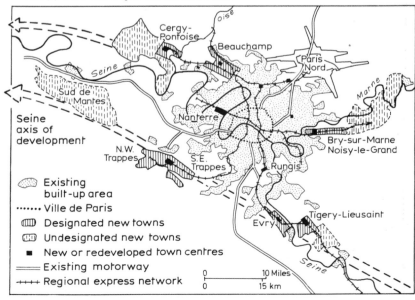

3.20(b) Cergy-Pontoise, France

industry when the only shop in the settlement belonged to the factory owner so the workers spent the wages he gave them in his shop. Very often they got into debt and became dependent on the owner.

These are some of the opinions of the inhabitants:

'Volkswagen is everywhere, you cannot get away from Volkswagen'

'I moved from a Volkswagen flat into my own house to get some independence from the company. I did not feel my life was my own'

'We like living in a modern new town. There are very good amenities and we do not mind if the nursery is owned and run by Volkswagen because the children like it'

'Wages are good and the accommodation is of a high standard.

located along the main lines of communication such as the valley of the Seine which forms an *axis* of development, the keystone of French regional planning policy.

Wolfsburg – a West German New Town

Wolfsburg owes its existence to the Volkswagen company. Begun just before the Second World War to produce the 'peoples' car' the factory now employs over 40 000, nearly all of whom live in Wolfsburg (fig. 3.21). This solves the problem of long distance commuting but accentuates the problem of dependence on one dominant industry.

Such is the hold the company exerts over the town that it is possible to be born in a Volkswagen hospital, live in a Volkswagen house, play in a Volkswagen park, be entertained in a Volkswagen cinema, and be buried in a Volkswagen cemetery. The company built the houses and flats and are the landlord to whom the tenants, their workers, pay rents. It is reminiscent of the early days of

3.21 Wolfsburg, West Germany

The company does not take advantage of its monopoly. If it was not for the factory physically dominating the town you would be unaware of its presence'

'It is difficult to make friends because the shift working from 05.30 to 14.00 hours and from 14.00 to 22.00 hours does not allow much social life and our friends all seem to be on the other shift'

1 Write an account of living in a 'One Industry Town' giving reasons why you would or would not like to live in one.

3.22 Shopping centres in England and Wales – and their importance

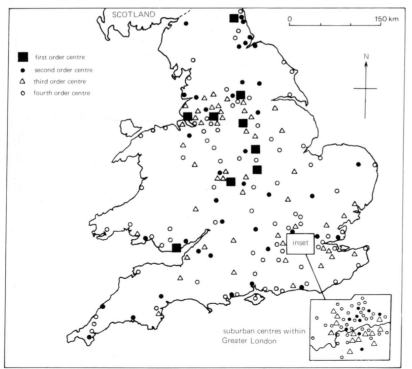

Hierarchy of settlements

The government conducted a survey of shopping centres in England and Wales and ranked the settlements into a hierarchy shown in fig. 3.22.

1 Count the number of first, second, third and fourth order centres.
2 Why are there fewer first order centres than any other, and more fourth order?
3 Using a compass, place the point on each of the first order centres and draw
  a a radius of 80 km. (50 miles) around each centre.
  b a radius of 160 km. (100 miles) around each centre.
4 Shade each of the radii a distinctive shading or colour.
5 Using an atlas, which areas are:
  a more than 80 km. from a first order shopping centre?
  b more than 160 km. from a first order shopping centre?
6 Explain why these areas are more than 160 km. from a first order shopping centre.
7 Turn back to fig. 3.22 and suggest how this hierarchy might be modified in the next 25 years.

It is possible to rank places by criteria other than shopping provision. Most frequently, places are ranked according to their populations. Fig. 3.23 shows the major towns and cities of England and Wales ranked according to their populations.

The places can be grouped according to their ranks. Fig. 3.23 has one such grouping but you might disagree and have more or less ranks. If so, put tracing paper over fig. 3.23 and draw diagonal lines to distinguish each rank. Write the rank alongside as in fig. 3.23.

A primate city

Whether you agreed with the division into ranks or not, it is clear that one place stands head and shoulders above every other place in the hierarchy, and that is London. Not only is central London a first ranking centre and has a far greater number and

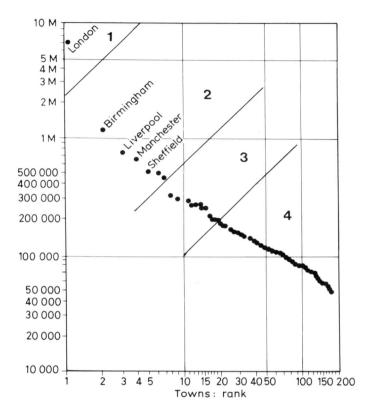

3.23 Population hierarchy in England and Wales

the extremes of the countries. Its commuting field is shown in fig. 3.13.. Besides housing the legal experts and the consultant surgeons and specialists, London is the origin of the national daily newspapers (fig. 3.24) which reach every corner of England and Wales; the centre of justice which could try a case from any part of the country; broadcasting and television stations transmit their national network programmes from here; it is the focus of entertain-

3.24 Transport and other facilities make London a primate city

variety of large stores than any other first ranking settlement in the country, it also houses a large number of second and other ranking centres within its boundaries. This can be largely explained by London's predominance in the population hierarchy. In addition London's *threshold population* is certainly drawn from all of the south-east of England which contains one third of the entire population of England and Wales – 17 out of 51 millions. But it can be argued that London's *sphere of influence* really covers the whole of England and Wales and its threshold population stretches to

ment with the leading cinemas and theatres and sports grounds staging cup finals and international matches. People come to London to get, see or hear things that they cannot get, see or hear in any other place in the country. A place which dominates to this extent is known as a *primate* city. This degree of *primacy* is not confined to this country nor to the Developed World. It is an even more significant characteristic of the majority of countries in the Developing World.

The Thames Barrage

'In 1236 the Thames overflowed its banks causing the marshes around Woolwich to be all at sea ... and in the great Palace of Westminster men did row with wherries in the midst of the Hall.' In 1663 Pepys wrote in his diary that the greatest tide that ever was remembered in England to have been in this river: all Whitehall was drowned! The floods of 1928 and 1953 claimed 314 lives between them.

One morning you might wake up to read in newspapers published outside London or hear on the radio the following news flash:

'At Westminster last night a wall of water swept over the Embankment surging into the House of Commons. The Underground from Westminster to the Temple became a watery grave for thousands travelling in rush hour trains. Nearly a million people over 60 square miles are under water; motorists drown in their cars; pedestrians too slow to get to safety.

Power, gas and water supplies are knocked out, telephone and teleprinter services cut, and factories, shops and offices disabled. Unknown numbers are still trapped in upper storeys. The whole nation is affected by the capital's paralysis. The cost of repairing the damage will run into thousands of millions of pounds.'

It is to prevent a catastrophe of this magnitude that the Greater London Council and the Government are financing the construction of a barrier in Woolwich Reach which will cost about £1000 million by the time it is completed. There will be four 200 m long and 20 m high steel gates and eight smaller ones. They will lie in concrete recesses in the bed of the Thames when not in use. When a flood warning is given the gates will swing upright cutting off all flow upriver. Walls will be raised downstream to protect against the return and rising water.

At the same time sirens will sound, public transport will halt, the Underground will be evacuated and storm doors closed. Radio One will broadcast continuous warnings and urge people to seek high ground immediately and await further instructions. The threat of such a catastrophe occurs only once a year although alerts are more frequent. But with Britain tilting south eastwards and south east England sinking at the rate of a metre a century (while north west Scotland rises by a similar amount) the frequency is predicted to increase to 10 alarms a year by 2030.

Despite these elaborate precautions it seems as if there could be serious loss of life because of people's *perception* of the danger. In a district in London, although 79 per cent had experienced the 1953 storm surge and suffered damage and they were still living in the same houses, only 2 per cent expected to suffer damage again. The majority knew of the protection to be afforded by the Thames Barrier and the raised walls but 48 per cent thought that this might serve to increase the risk. Most considered their previous experience to be unique and a 'freak' and it would never occur again in their lifetime. This emphasises how much the public needs to be educated about the dangers and how authorities and planners have got to take cognisance of people's perception of and their attitudes to *hazards*.

The Thames Barrage is designed to protect London and avert national disaster, not to increase the water supply of the capital.

1 Is there a danger of flooding in the area where you live? If so, conduct a survey to assess the inhabitants' perception of the danger.

2 Is there any danger from any other hazard in your area such as a moving coal tip, as the one which engulfed the school in Aberfan, South Wales burying children alive?

3 Why do you think people 'ignore' real dangers such as those cited above?

4 **Either** imagine you were rescued from the flood and were being interviewed in hospital for an eye witness account. Write your story.

Or imagine that a similar disaster has befallen the place where you live. What news flash might appear in the local newspaper or be broadcast over local radio?

4 The Structure of Cities

Central Business District

Look at fig. 4.1. It summarises the main features of the structure of cities. It is like the view you get if you slice a cake in half.

The 'three-dimensional' model in fig. 4.2 expresses the pre-eminence of the main road. But it does more. It shows that the values are high at junctions, and that the roads lead to one zone with land values far higher than along any of the roads.

1 Where are the land values highest?

2 Do land values increase sharply or decrease sharply as you move away from the main roads?

Why are values so high in this zone? Many local authorities have asked the same question and conducted surveys to find the answer. A survey conducted in Norwich found that over half the employed population work in the central zone of the city, over half of the population visit the city centre at least once a week for shopping, while over a third visit it for other purposes such

4.1 Land use zones in a city

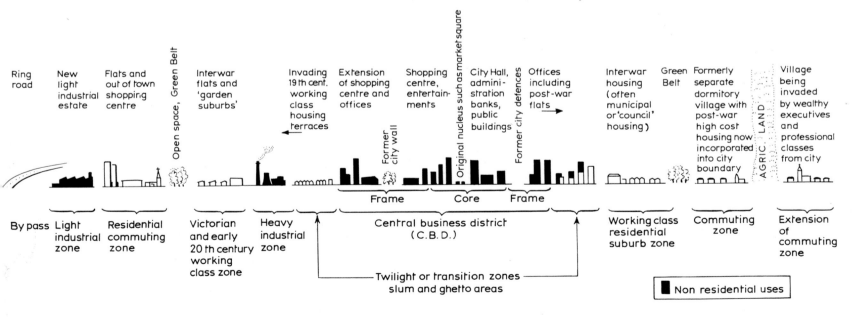

as recreation, professional services or educational purposes. Some of these journeys will be for a single purpose only but others will combine two or more purposes on one trip. This pattern is typical of most towns and cities and emphasises the pull of the centre.

A high price to pay for accessibility – the peak land value intersection

Shops and offices need to locate where they can attract the most custom. Most custom will be obtained where there are the greatest pedestrian flows. Pedestrians tend to flow in straight lines so the best opportunities for soliciting custom will be in the main streets of the central business district. Since the main streets occupy a limited zone and many businesses will want to locate along them, competition for the sites will be severe and drive up the prices of these desirable sites. Businesses are prepared to pay as high a price as they can afford as long as they can still make a profit. Competition forces the costs of the most accessible point in the Central Business District (CBD) to a peak where it is higher than anywhere else in the settlement. This point is known as the *peak land value intersection* (fig. 4.2).

In order to obtain the high returns needed to pay the high price of locating in the best positions in the CBD, firms have to make the fullest use of their premises. This means building *vertically* and using every floor to its best advantage. As a result of these pressures a common distinguishing feature of many of our town centres are tower blocks or 'skyscrapers' with shops on the lower floors, and offices and, occasionally, residential uses above.

The one use that has been largely driven out of the central zone is residential. The busiest thoroughfares were once fashionable residential areas, being most accessible to the few shops that were available. Today the land in the central zone is so expensive that very few people can afford to live there.

As a result the central area of a town has:

1 The greatest concentration of shops, offices, cultural facilities, entertainment, transport termini, and administrative buildings.

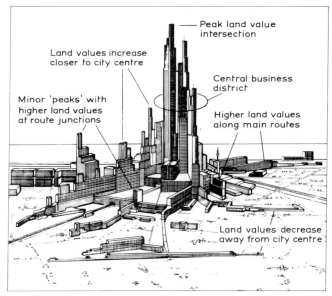

4.2 A model of land values in Copenhagen

2 A significant share of wholesaling and industrial functions.
3 The focus of transportation routes.
4 The busiest telephone network of any part of the town.
5 The greatest concentration of people and vehicles.
6 The highest land values.
7 Most of the highest buildings.

Death of the High Street – birth of the hypermarket

Some people have forecast that congestion will choke the High Streets in the CBD to death and encourage the development of alternative forms of retailing such as out-of-town hypermarkets. Fig. 4.3 shows this new form of retailing which has revolutionised shopping patterns in North America. However the photograph was not taken in North America nor continental Europe but in Irlam outside Manchester.

4.3 A hypermarket near Manchester

4.4 The hypermarket debate

The argument in favour of hypermarkets

1. They help relieve traffic problems in congested city centres.

2. Town centres cannot provide sufficient car parking space.

3. Because of the huge scale of the operation and the low rents resulting from an out-of-town location they can cut prices.

4. There is little evidence that they have a detrimental effect on adjacent town centres.

5. There is evidence that even in the absence of competi-

The argument against hypermarkets

They tend to create serious traffic problems.

The vast car parking area and the meccano-like construction are aesthetically displeasing and difficult to disguise.

There is no guarantee that once established they will adhere to their policy of cutting prices, especially when they could have a monopoly.

They could seriously undermine the central area retailing and other business activities.

Deterioration in central area retailing threatens the continued

tive new forms of retailing some town centres are experiencing a decline in trade because they have not adjusted to increased car traffic and changing shopping habits.

6. Ample free parking and locations away from congested town centres make them highly accessible to car-borne shoppers.

7. Their range of goods enables basic shopping needs for the entire week to be purchased in one place on one shopping expedition in warm sheltered surroundings.

8. The unsightliness of some hypermarkets on the continent can be avoided by sensible landscaping, especially to hide the car parking.

9. They stimulate healthy competition amongst other forms of retailing.

10. If hypermarkets are banned then less easy-to-control retailing outlets will proliferate.

11. Hypermarkets are firmly established in the USA and in many continental countries, especially Western Germany and France and have proved their success.

existence of the central area as a social and cultural focus of urban life.

A significant proportion of the population (25 per cent) are and will remain non-car owners.

The increasing cost of petrol will deter people from using the car for shopping expeditions.

The large site required by a hypermarket offsets the lower cost of land in an out-of-town location so there is no saving in rent to pass on to the customer.

The most vigorous opposition comes from the local chamber of trade and commerce who fear for their livelihood.

Extensive redevelopment and pedestrianisation of many town centres is making them once again attractive shopping environments and there is no necessity for out-of-town shopping centres.

You cannot equate Britain with other experience since our town centres have had a different history and have occupied different roles from those in North America, Western Germany or France.

1 Why is this new form of retailing called a hypermarket?
2 What is revolutionary about it?

In 1960 there were no hypermarkets on the continent of Western Europe. Now there are more than 500, each with a minimum of 50 000 sq. metres of selling space besides the huge car parking space surrounding the store. Planning authorities in Britain have been more cautious, and only a few applications have been granted.

Some of the points considered in the debate are listed on page 41.

4.5 A pedestrian precinct in Wales

Pedestrianisation

The increasing volume of vehicle traffic converging on city centres has caused congestion and delays, and threatens the very accessibility which gave rise to and perpetuated the central business district. In addition, moving about the centre becomes hazardous and dangerous. Pollution and damage to the older buildings is caused by the constant vibration of traffic and the whole city centre environment deteriorates. Some city centres have responded to the challenge of hypermarkets by removing traffic from the main shopping streets,

paving the area, furnishing it with 'street furniture', planting 'instant' trees and converting it into pedestrians-only precincts. A safe, pleasant environment has thus been created (fig. 4.5).

The conversion of city centres to pedestrian precincts has gathered momentum and they are regarded as the hallmark of progressive planning. Therein lies the danger. Only the streets which are pedestrianised are upgraded. Here rents rise and only the department stores and branches of national multiple stores can afford the rents. The local and small shopkeepers have to quit the area and it loses its character, assuming a uniformity from place to place. The latest trend is to transform the entire city centre, encompassing not just the shopping streets but the offices, churches, and other public buildings. Evidence shows that this encourages people to revitalise the city centre as a social meeting place as well as a business centre.

1 In your opinion does pedestrianisation
   a improve the environment
   b cause a deterioration in the environment
   c increase or reduce the potential of the town centre as a shopping focus
   d make it safer or less safe?

Location of stores in CBD

Goods can be divided into three categories:—
1 *Convenience* goods which people require daily such as bread, butter, eggs.
2 *Comparison* goods such as clothing which people require less frequently but which call for greater inspection and comparison before purchasing.
3 *Luxury* or *Specialist* goods such as jewellery or a musical instrument are required infrequently. They are expensive and are purchased comparatively rarely. They often include purchases made 'once in a lifetime' such as a piano or expensive ring.

The frequency of purchases and the cost of the goods affects where these stores are located in the CBD. The other factor is

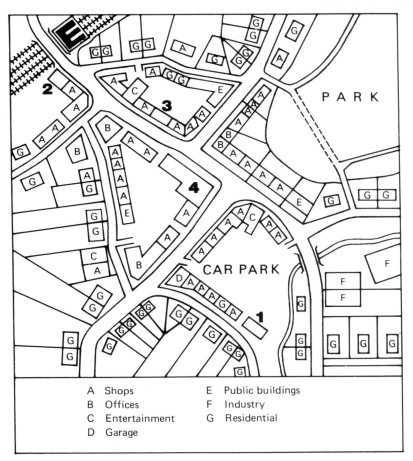

4.6 Business in a town centre

Legend:
A Shops
B Offices
C Entertainment
D Garage
E Public buildings
F Industry
G Residential

these extremes are the comparison and the luxury goods. The comparison goods are usually closer to the peak land value intersection than the luxury stores because they sell fairly expensive goods more often than the luxury stores whose total sales are not as high, and therefore they can afford the higher costs involved in locating close to the centre of the core of the CBD.

The effect of this sorting out process depends on the ability to pay the rents and rates demanded in different parts of the CBD.

1 Look at fig. 4.6. Four shops have become vacant in this town centre. Choose from the following list the four businesses most likely to locate in each of the vacant lots. Match the appropriate letter attached to each store with the number in the vacant space.

A – Estate Agent   E – Motor Accessories
B – Toy Shop   F – Hairdresser
C – Footwear   G – Bank
D – Travel Agent   H – Supermarket

Getting together

When you buy a house you need the services of an estate agent, a solicitor, a building society and an insurance company. It is an advantage if the business can be executed efficiently and quickly. This is easiest if these locate near to each other. There is a direct *link* involved between them and therefore they need to be *accessible* to one another.

Very often similar shops *cluster* together. These are often complementary to each other. If a customer cannot find what she wants in one shop it is easy to slip in next door to compare the ranges.

Much of an antique dealer's trade derives from other antique dealers because of a tendency towards *specialisation* within the trade. Antique dealers often purchase the contents of whole houses although they are only interested in a small fraction of the purchase. But they know that they will be able to deal with their neighbours easily and dispose of the surplus because they specialise in some

the cost of the site in rents and rates. Department stores contain all categories of goods under one roof and they can afford the highest rents and rates. The stores which sell the relatively inexpensive day to day convenience goods cannot afford high rents and rates and locate on the edges of the CBD. Somewhere in between

other branch of the trade. Picturesque historic enclaves such as The Lanes in Brighton foster the image and encourage *clustering* of like or complementary functions, especially in places with a well developed tourist industry.

A final example is the dispensing chemist whose best location is next door or immediately opposite the doctor's surgery. No doubt you can think of many other examples. In the very largest cities such as London there are distinct quarters or zones where similar activities congregate, as shown in fig. 4.7.

4.7 Zones of land use: central London

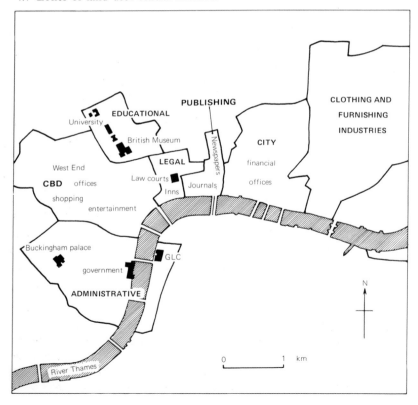

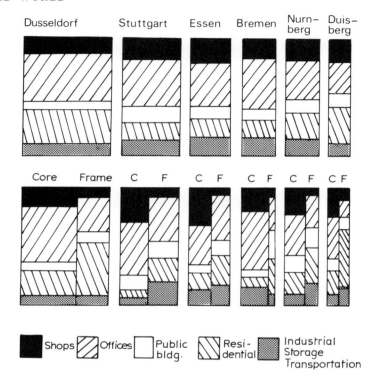

4.8 Use of land in the central business district in six West German cities

Defining the CBD

Despite its apparent uniformity it is possible to distinguish between the heart or *core* of the CBD and the margin or *frame* which surrounds it on the basis of the floor area devoted to the major uses of land (fig. 4.8). It is also possible to differentiate the CBD from other zones by the higher rateable values on properties, and other methods such as a rateable index which divides the area occupied by a property by the rateable value. None are entirely satisfactory and depending on which you use you would draw a different boundary. The greatest uncertainty occurs around the edges of the CBD.

Twilight zone

Studies of the margins of a large number of central business districts have resulted in the following list of typical uses of land:

Professional Services such as Solicitors, Auctioneers and Doctors' and Dentists' surgeries.
Administrative Buildings such as Town Halls.
Other Public Buildings such as Churches.
Warehouses.
Wholesale establishments.
Auction Rooms.
Garages.
Public Transport terminals such as Bus and Railway Stations.

Many of these are found in the zone immediately adjoining the CBD, often called the 'Twilight zone'.

1 Walk around the centre of your own town and mark on a 1 : 2500 or a 1 : 10 560 map the uses of all the buildings. Distinguish between the part which has mainly shops, offices including banks, estate agents and building societies, hotels, cafes and cinemas, which are typical land uses found in the CBD, and the part which has the uses listed above.
2 Study fig. 4.9 which shows the CBD of Crawley, a New Town 25 miles south of London. Describe and explain the location of
  i) transport facilities including car parks
  ii) public services
  iii) leisure amenities
  iv) the shopping precinct
3 What evidence is there to suggest that the new town centre was built near to or adjacent to the former centre of Crawley?

Inner suburbs: poverty in the midst of plenty

Lying between the twilight zone and the outer suburbs are the inner suburbs. About 100 years ago they were themselves the outer

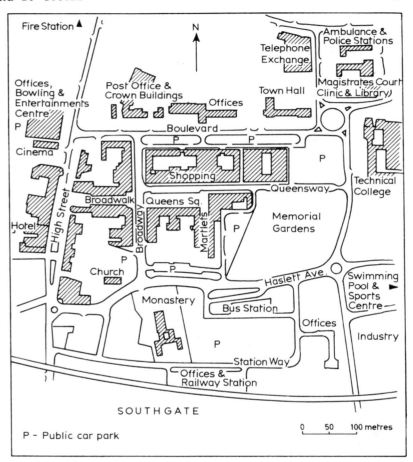

4.9 Crawley New Town: central business district

suburbs on the edges of the cities, but the expansion of the cities has gone far beyond their limits. A large majority of the houses were built without baths, with outside lavatories and with cold water only. Despite house improvement grants the majority still lack what the majority of the population consider basic amenities.

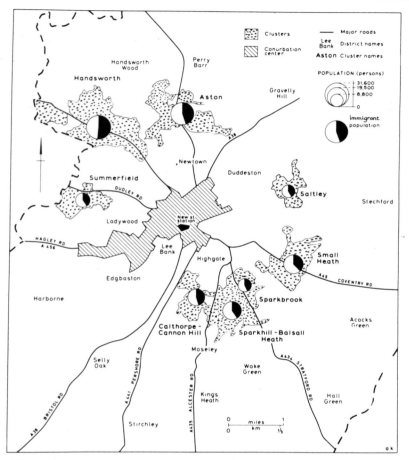

4.10 Concentration of immigrant population in Birmingham

A survey in Britain identified 126 areas of such *multiple deprivation*. Nearly all were in the largest cities, with the highest number in Glasgow.

As the inhabitants of these once fashionable areas left for the newer suburbs, they were replaced by newcomers seeking a foothold in the city. In cities such as New York, Chicago, Bradford and Birmingham it is possible to trace successive waves of *invasion* and *succession* by different ethnic groups including Central Europeans, Irish, Asian, West Indian and Puerto Ricans all trying to escape from unemployment, poverty and famine in their own countries. (fig. 4.10). The initial groups found jobs nearby and moved on. However, more recent immigrants have found it difficult to find employment as industry has been moved to the suburbs, and the jobs that are available demand skills which not all of them possess. Poverty means that maintenance of properties is neglected and dilapidation of the urban fabric is rapid. Such areas can become *ghettos* housing the groups trapped by their inability to flee wretched living conditions which become breeding grounds of vandalism, crime and violence. A number of steps can be taken to improve them. They can be cleared and redeveloped, they can be renewed or they can be rehabilitated. These are all expensive solutions and choices have to be made. However governments as in Britain have realised that resources which formerly went to prestigious projects such as new and expanded towns must be redistributed to resurrect the inner parts of cities. It must not be imagined that every part of the inner city exhibits these symptoms because it does contain high class areas such as Mayfair and Chelsea in London. However, generally the inner city is in decline and has a higher incidence of problems such as low incomes, high rents, overcrowding, high child densities, long housing waiting lists, dependence on welfare services, homelessness, squatting and racial friction. Replanning must be comprehensive taking account of both the physical aspects such as design and lay out and the social aspects.

Density and standards of housing

Fig. 4.11 shows the relationship between the density of population in different types of housing and the satisfaction or dissatisfaction of the residents with this type of housing.

1 Which type of housing causes least dissatisfaction?
2 Which type of housing causes most dissatisfaction?

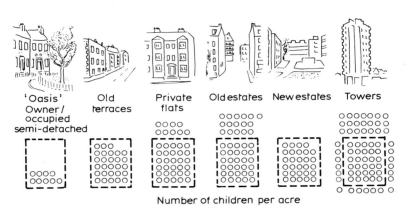

Number of children per acre

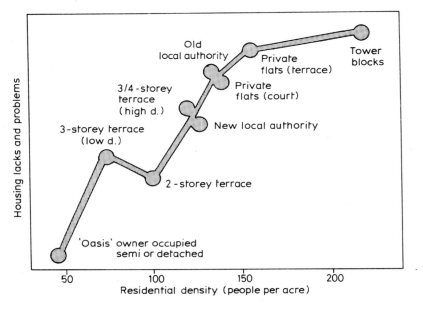

4.11 Density and dissatisfaction with housing in London borough of Lambeth

3 Why are there these differences in the level of satisfaction between these two types of housing?

4 Which of these types are most common in inner city areas? And which is most common in the suburbs?

Urban renewal

Suburbia, Expanded Towns and New Towns are all associated with the desire to improve the 'quality of life'. In many instances this has been accomplished at the expense of the inner city which becomes run down and decayed. Recently attention has been directed at this problem and the process of rejuvenating this area has been termed *urban renewal*. The problem is not new, what is new is an increasing public awareness of the problems.

Legislation giving compulsory purchase powers and social crusading zeal has put Britain in the forefront of renewal programmes. An additional reason is that much of the property in the inner areas of towns and cities was built on 99 year leases which were 'falling in' in the post-World War II period (post 1945) making whole areas ripe for redevelopment. Other legislation provides assistance for leaseholders to purchase their properties and provides grants for the rehabilitation of properties up to contemporary standards.

Moving out

Oldham lies on the western slopes of the Pennines, 10 kilometres from the centre of Manchester. It grew rapidly in the nineteenth century when it became an important cotton spinning town. St. Mary's is a district near the town centre which had deteriorated into a *slum*. A slum has similar physical characteristics and problems to the ghetto but it houses a closer knit community which has lived in that neighbourhood for many generations. By the 1960s the houses were considered uninhabitable because only 4 per cent had fixed baths and only 17 per cent had running hot water. The cost of 'regenerating' these properties to contemporary standards would have been too high. It was decided to launch a 'slum clearance' programme and 'renew' the central area. The 2370 people living in St Mary's had to be rehoused. Fig. 4.12 shows where they went.

1 Who rehoused most of the people who formerly lived in St Mary's?

2 Where were the majority rehoused?

3 What disadvantages would they have found in their new houses compared with living in St Mary's?

4 What do you think would happen to the community spirit that had been built up in St Mary's when the population had to move out? Why?

5 Where did most of those who could afford to buy their own homes decide to live? Can you suggest reasons why they settled here rather than further out from the town centre?

Rehabilitation

Is it possible to prevent the break-up of communities like St Mary's? It is possible to improve or rehabilitate the houses. A government publication says 'an area is only worth improving if the property is sound, if the residents still care about it, and if there is space to make improvements'.

Armed with these guidelines local authorities investigate the viability of various lines of action before proceeding with demolition and renewal or rehabilitation. What would you do if you were confronted with a locality like this in Nelson? (fig. 4.13).

Bearing in mind that the council would have to rehouse the inhabitants if they demolished the properties, and the cost of each new unit, they considered that this area was a social asset and worth preserving. The houses were a convenient size and easy to run, and near to shops and transport routes. There was evidence of a neighbourliness often found in long-established working class districts. House prices were low enough for most of the residents to attain the ambition of owning their own house without financial hardship. Nearly all of these advantages would be lost if the area was cleared and the residents rehoused elsewhere.

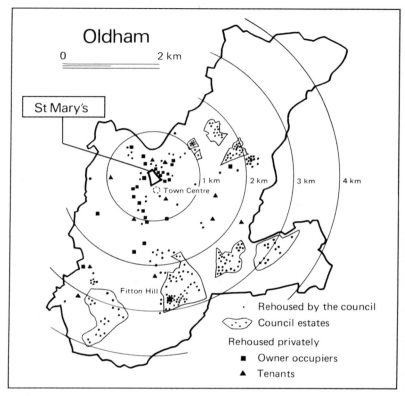

4.12 Slum clearance area: dispersal of former residents

Moving back – The Barbican

In 1851 the City of London was densely populated by nearly 128 000 people. Now less than 5000 inhabit the area. In the former garment manufacturing sector of the city, Cripplegate, where 14 000 had lived a century earlier, only 48 were left in 1951. World War II had cleared most of the buildings. Was this space to be put to its most lucrative use for office development or was

4.13 Rehabilitation of an inner city area

1 The retention of St Giles Church, Cripplegate, which becomes the focus of the development, and parts of the Roman Wall as 'landscape' features.

2 A generous provision of open space including gardens and a lake covering nearly 1 hectare.

3 The separation of vehicles and pedestrians is achieved by a raised walkway 6 metres above the present road linking various parts of the development and the development to other parts of the City, thereby creating an artificial 'ground level'.

4 New accommodation for the City of London School for Girls and the Guildhall School of Music and Drama.

5 A theatre designed to house the Royal Shakespeare Company.

6 A concert hall to be the permanent home of the London Symphony Orchestra.

7 An arts centre which incorporates the theatre and concert hall which can also double up as a conference centre for 2000 delegates.

8 Shops, restaurants, library and public houses.

1 Which facilities are designed for the Barbican residents only?

2 Which facilities are designed to make the Barbican the City's cultural as well as a commercial focus?

3 Do you think that the residents of the Barbican will be similar to those living there before World War II? Why?

4 Write a paragraph to say to what extent you consider this to be a successful attempt to bring people back to live in the city centres.

Rebirth of St Katharine's Dock 1828–1968

Look at fig. 4.15. This is hardly how one would expect a 150-year-old dock to look! In 1967 St Katharine's Dock in the Port of London died, its purpose in life complete. The shipping which had led Thomas Telford to clear 1000 slum houses and develop the 10 ha (hectare) site as a dock in 1827 had long since deserted her and moved downstream as they grew in size. Then in 1969 the Greater London Council bought the derelict site from the Port of London authority, and breathed life back into the dock.

the opportunity going to be taken to breathe new life into a city of 'cats and caretakers'?

Fig. 4.14 gives the answer. The Barbican scheme promoted by the Corporation of the City of London aims to accommodate 6500 people in just over 2000 flats and maisonettes on a 15.2 hectare site. The remaining 25 hectares of the site are being developed for offices. It has a number of individual features:

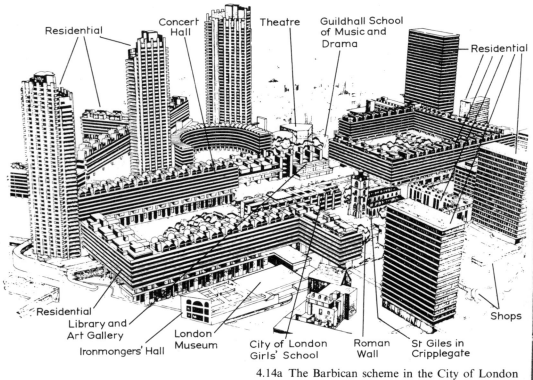

4.14a The Barbican scheme in the City of London

4.14b A new block of flats contrasted with St Giles, Cripplegate

It had purchased a prime site next to the Tower of London and overlooking Tower Bridge, but what could be accomplished with a stretch of water and some forbidding, dilapidated warehouses? Fig. 4.15 indicates the transformation that has already taken place, and by its scheduled completion in 1985 it is planned to include:

1 2 or 3 hotels. The Tower Hotel with 826 rooms and 3 restaurants is already open and the other, Europe House, with conference and exhibition facilities, is in the process of construction.

2 300 units of local authority housing.

3 378 units of private housing, including some in renovated warehouses such as Ivory House, the name a reminder of former cargoes landed here.

4 A British export centre.

5 A World Trade Centre.

6 A theatre, shops, cafés, a chapel, and a primary school.

4.15 St Katharine's Dock, London

7 A $2\frac{1}{2}$ ha yacht basin with 240 berths and attractions such as the Old Nore Lightship already in position and some old Thames sailing barges as a testimony to the past.

8 One warehouse when stripped was discovered to be a timber framed construction and is being preserved and moved 100 metres before conversion into a tavern and restaurant.

1 Write a paragraph with the heading: 'St. Katharine's Dock: a Major Conservation Achievement', or, 'St. Katharine's Dock: a Bold and Imaginative Urban Renewal Scheme', or 'St Katharine's Dock: an Unnecessary Extravagance'.

A 'model' of suburban development

Suburban development is the product of the natural process of the growth of towns to accommodate the increasing population. It is not a deliberate creation built on virgin territory separated from the parent settlement but a continuation of the built up area occupying the fields which were once the limit of the settlement. The spread of suburbia was assisted and in many instances promoted by the railway and underground companies. Enticing advertising (Fig. 3.12) encouraged those who could afford to leave the inner city areas and realise their dream of a 'castle' and a garden in the countryside. This movement gathered momentum in the later years of the nineteenth century and throughout the twentieth century. As transport services became faster and more efficient the suburbs eat up more and more of the countryside and early suburbs were engulfed in the onward march of bricks and mortar.

This process is summarised in the 'model' of suburban development in fig. 4.16.

1 What are the characteristics of the area shown in fig. 4.16 before the advent of suburbanisation?

2 What evidence is there in fig. 4.16 to indicate that the authority governing the area had changed?

3 It was the individual family who pioneered the way into the suburbs. They were followed by the speculative builder who first developed market gardens and single fields and later entire farms with houses that were often identical or very similar in appearance. Finally private and public bodies combined to induce a mass movement of people to fill up the suburb. Why was there little resistance to the acquisition of this type of land for building purposes?

4 What evidence is there in fig. 4.16 to suggest that residential segregation was beginning to affect the suburb? What was the basis of this residential segregation?

5 What amenities were provided for the expanding population? Were they adequate? Why?

6 What happened to the commercial area as the suburb matured? Why?

7 What restriction is there on further growth of the suburb?

8 If you live in a suburb or there is one near you, try and reconstruct the growth of your suburb along the same lines as in fig. 4.16. Did it develop in a similar way and at a similar

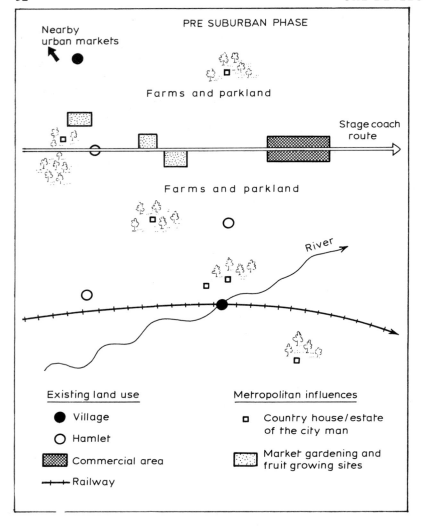

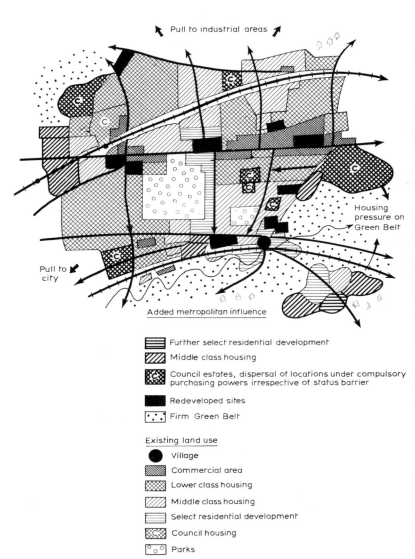

4.16 A model of suburban development

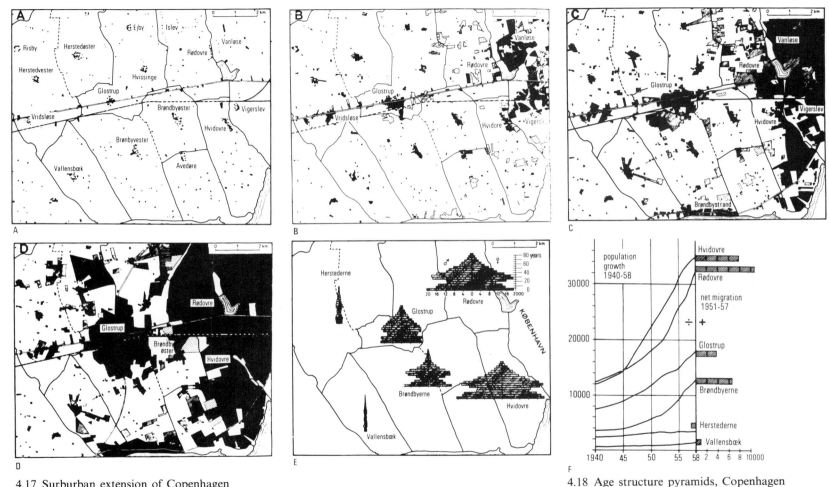

4.17 Surburban extension of Copenhagen

4.18 Age structure pyramids, Copenhagen

time? Explain differences and similarities between the development of your suburb and the model.

9 Put yourself in the place of an inhabitant who has lived in the same house for over 50 years, during which time his village has been transformed into an integral part of the urban complex.

Describe what has happened and what you feel about the changes.

10 Given the choice, would you prefer to live in a downtown ghetto or in a suburb on the outskirts of the settlement? Why?

Gobbling up the market gardens

The suburban extension of Copenhagen is demonstrated in fig. 4.17.

1 In 1860 the Danish capital had not impinged on the villages at all. They remained unaffected by the expansion of Copenhagen, which was surrounded by market gardens.
2 In 1910 suburban development was penetrating into those parishes nearest Copenhagen. A small town had grown up around the railway community of Glostrup. Market gardening was gradually spreading westwards.
3 By 1930 the parishes of Vanlose and Vigerslev nearest Copenhagen were rapidly filling up and expansion had continued into the neighbouring parishes of Rodovre and Hvidovre. Market gardens and week-end gardens (allotments) were being pushed further west.
4 By 1960 the four parishes Vanlose, Vigerslev, Rodovre and Hvidovre were fully developed suburbs. Glostrup had grown rapidly with the acquisition of extensive industrial areas. Market gardens had spread even further west in the face of the advancing human tide and increasingly larger areas were devoted to such uses.

1 Draw a rectangle in your note books. Put the present year as your title and add the extent by which Copenhagen has extended even further westwards between 1960 and the present. If you turn to page 56 you will see that the seemingly inevitable expansion has been halted. How has this been done?

The age structure pyramids (fig. 4.18) give an insight into the demographic characteristics of the suburban population. The swelling in the middle of the pyramids of Rodovre, Hvidovre, and Brondbyerne indicate young married couples. The swelling at the bottom of the pyramids indicate the children of these parents. The pyramid for Glostrup does not have such pronounced swellings at the bottom and in the middle because the newcomers have been added to an existing town. As a result there are more older people in Glostrup so that the pyramid does not have such a sharp apex as in Brondbyerne. The pyramids for the rural parishes of Herstederne and Vallensboek which are still in the pre-suburban phase have constrictions in the middle and at the bottom. This reflects the emigration of young adults (probably they have moved into Copenhagen) and the absence of the children who would have been born to them had they remained in the parishes.

There are other characteristics of suburbia which cannot be shown on such pyramids: that most suburban dwellers are married, own their own houses, have more than one car, and many of the wives do not work.

City regions – planning to control urban growth

Cities have a profound effect on their surrounding area. Primate and capital cities such as London and Paris organise their surrounding area as a *tributary zone* functioning as one 'region' or 'system' with the city. When this occurs planning has to be undertaken for the whole region because decisions affect the operation of the entire system. Figs. 4.19, 4.20 and 4.21 illustrate some of the plans produced for some of these regions.

London's Green Belt

Steps have been taken at intervals throughout this century to try to control *urban sprawl*. Plans to contain the growth of urban England did not become really effective until the Green Belt Act 1938 gave them legal status. This 'cordon sanitaire' was extended and the powers of control tightened in the Greater London Plan 1944. Action of this kind was necessary to prevent an *English megalopolis* stretching along the London to Merseyside *axis* via Birmingham and the Midlands. Nevertheless, planning cannot be solely restrictive and it has to allow for the expected increase of 4–5 million inhabitants by 2000 AD (fig. 4.19).

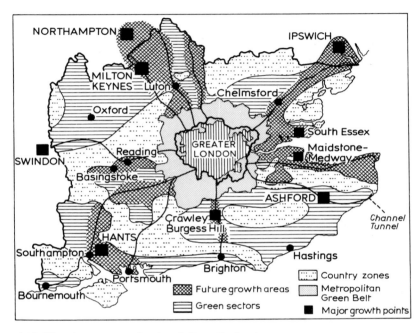

4.19 Planning strategy for South-East England

In order to accommodate this increase and provide the houses, jobs and other services necessary to support it, a series of studies have been undertaken and plans produced. In 1964 there was 'The South-east Study, 1961–81', in 1967 – 'A Strategy for the South-east', and in 1970 – 'Strategic Plan for the South-east', reviewed in 1976 and 1978.

1 Why was the London-Merseyside axis the most likely location for a megalopolis in Britain?
2 About how wide is the Green Belt at its narrowest and widest points? Use an atlas and try to explain the different widths.
3 Use the atlas to discover the names and type of land designated as Green Sectors and Country Zones. Why have the planners chosen to preserve these areas as 'lungs' for the urban population rather than others?
4 What two principal strategies have the planners adopted to accommodate the anticipated growth in population? Use fig. 4.19 and an atlas and suggest characteristics which these growth areas have in common and which probably contributed to their selection in preference to other potential growth areas.
5 Try and find other examples of Green Belts and compare them with the metropolitan Green Belt around London.

Copenhagen Finger Plan

The population of the city of Copenhagen reached a peak in the 1950s of 777 000, since when it has declined steadily to its present figure of 600 000. At the same time there has been a corresponding increase in the population of the Copenhagen conurbation to 1 700 000. This is over 33 per cent of the entire Danish population and makes Copenhagen a good example of a primate city since the second city in Denmark, Aarhus, only has a population of 250 000.

This process has produced the familiar problem of bridging the gap between home and work. In 1949 an Urban Restriction on Extension Act was passed which accepted an inner zone where urbanisation was acceptable and an outer zone which was to be preserved for agricultural and recreational purposes. It has proved an efficient means of regulating growth and preventing unsightly urban sprawl. But sprawl occurred just beyond the controlled area about 15 miles from Copenhagen. A voluntary cooperation of municipalities around the capital agreed to extend the regulations to include their areas. In 1969 the principle of zoning was extended to the whole of Denmark by law and the whole country is now divided into urban, and rural or green zones. Since 1967 the voluntary cooperation assumed a more permanent form as a Regional Planning Advisory Council but still only in an advisory capacity. Negotiations are now proceeding to establish a Metropolitan Council for the whole *city-region* dominated by Copenhagen and so

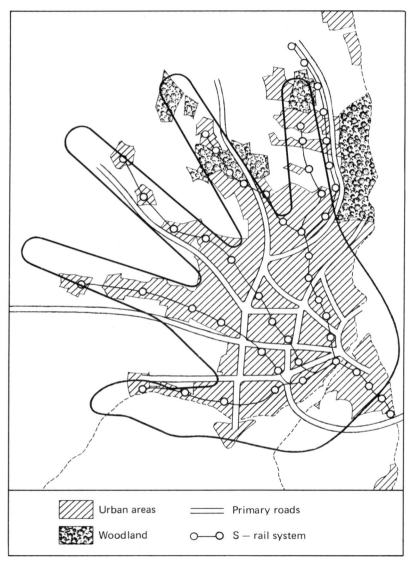

4.20 The Copenhagen 'Finger Plan'

Urban areas   ——— Primary roads

Woodland   ○—○ S – rail system

confirm the reality of the existence of a distinctive region controlled by the capital.

Despite the absence of a body with executive powers the municipalities agreed to a regional plan. The plan they adopted is shown in fig. 4.20 and is known as the 'Finger Plan'.

1 Why is it known as the 'Finger Plan'?
2 How are the planners encouraging development along these fingers?
3 What are the 'green wedges' being used for?
4 How is through traffic being kept out of the old city core?

The main principles underlying the 'Finger Plan' were:

1 to preserve the urban-rural separation.
2 to facilitate the journey to work and to limit it to 45 mins.
3 to introduce a three-tier provision of public transport
  a the existing trams in the inner city having a 45 min. range of 8–10 km.
  b the S-Bahn or suburban electric railway services with a 45 min. range of 15–20 km.
  c fast through or limited stop trains with a 45 min. range of 30–40 km.
4 not to carve up the historic core with urban motorways and attract cars into the centre but to provide ring roads to keep traffic away from the centre.
5 to keep goods terminals for trains and lorries outside the central area and along the ring roads.

Randstad, Holland

A part of the world where more than one conurbation has grown to *coalesce* with others is the Netherlands. It is the most densely populated country in Western Europe and one of the most densely populated in the world. There is an average of 389 people on

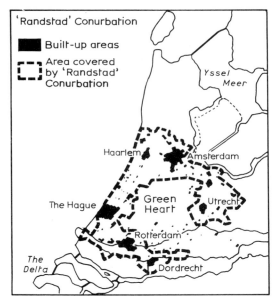

4.21 Ranstad Conurbation, Holland

every square kilometre (compared with an average of 344 per sq. km. in England) but in the urban areas of Randstad (fig. 4.21) the average rises to 2563 persons per sq. km. Both Amsterdam and Rotterdam have more than 1 million population and The Hague and Utrecht more than half a million each. The stages by which they have merged are depicted in fig. 4.22.

1 Had any of the major urban areas actually joined each other by 1930? by 1947? by 1960? by 1980?
2 In which directions did the various stages of growth occur?
3 Why did expansion take place in this part of the Netherlands and in these urban areas and not in other parts?
4 For example why was there relatively little expansion of Amsterdam and Haarlem northwards? Turn to page 58 and see if there is any likelihood of expansion in this direction in the future?

5 Write a few sentences to explain why this area is called Randstad or 'Ring City' Holland.
6 Write a few sentences to explain why another name for the same area is 'Greenheart Metropolis'.
7 If the region was stretched out similar to the urban area along the eastern seaboard of the USA how far would it be from end to end? Is this distance greater or less than the extent of Megalopolis USA? Is the density of population greater or less in this region than in Megalopolis USA?
8 Remembering that Randstad occupies about 20 per cent of the surface of the Netherlands and houses nearly half (6 million) of the total population at present and is expected to have to accommodate a further 2 million by 2000 AD, write a few sentences to say whether it would be correct to call this area 'Megalopolis Netherlands'.

The Dutch authorities have recognised the danger that the 'green heart' might disappear as a result of urban pressures and that this area could become a completely urbanised zone stretching from the North Sea coast to Arnhem and Utrecht in the east. In 1966 they published a Report of the Physical Planning of the Netherlands which proposed:

1 Stemming the 'drift to the west' of the country by encouraging industries to locate in towns such as Groningen and Arnhem. In this way 'counter magnets' would be developed to reduce the attraction of the west.
2 Creating 'buffer zones' of green countryside at least 4 km wide between each of the cities in Randstad.
3 Preserving the intensive farmlands of the 'green heart', such as the glasshouse districts, from encroachment.
4 Limiting future urban development along 'radial wedges' on main transport routes and retaining agricultural land to separate the urban wedges. One such wedge would connect Amsterdam and Lelystad (fig. 4.23), capital of the newly reclaimed Southern Flevoland polder, and another Rotterdam and Haringvliet as a consequence of the Delta Reclamation Scheme.

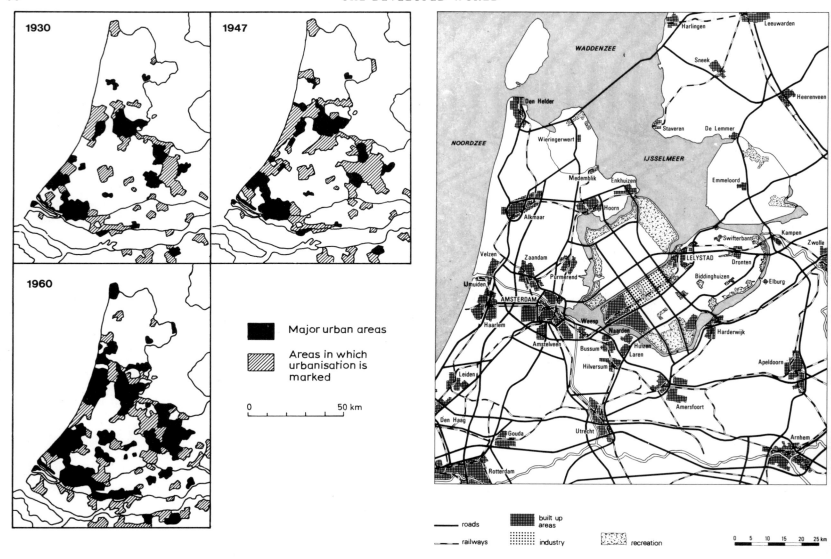

4.22 Process of merging of major cities, Randstad, Holland

4.23 The radial wedge connecting Amsterdam and Lelystad

5 Transport, Communications and Services

Motorways

The development of motorways began in Germany with the Auto-bahn in the 1930s as part of a military strategy. The idea was copied in the United States with their turnpikes and freeways. The first stretch of motorway in Britain, the Preston bypass, was not opened until 1958. In the early 1970s the target of 1000 miles (1600 km.) was reached. The pace of motorway building has slowed down but as fig. 5.1 shows there are still important gaps in the network.

1 Describe the pattern of motorways in Britain.
2 List the main gaps that remain to be closed.
3 Why do the motorways follow the routes that they do?
4 Explain the absence of motorways from certain parts of the country. (fig. 5.1 will help)
5 Why has it been necessary in some instances for motorways to duplicate existing trunk roads?
6 Give reasons why the pace of motorway construction is slowing down.
7 Why was it necessary to build motorways in the first place?
8 What advantages do motorways offer?
9 What particular dangers do motorways exaggerate?

Some of the gaps remain because of disagreement over the line the motorway is to take. Many long delays have occurred because of opposition to planned routes. These conflicts are sharpened when scenery is threatened.

Bridging the gap

Look at fig. 5.1a and an atlas:

1 How many miles does the Severn Bridge save on the journey between Bristol and Newport?
2 Besides the Severn estuary what other obstacles might have lengthened the travel time between Bristol and Newport before the bridge was opened?

It is interesting to compare what might happen as a result of the opening of the Humber bridge with what has happened as a result of the opening of the Severn bridge in 1966. This extract from a local newspaper tells its own tale:

'The Severn bridge is far from being simply an engineering miracle – to the ailing economy of South Wales it has been a Messiah. In the glow that flickered from the dying embers of the era when Coal was King one crucial conclusion was clearly visible – Wales's chance of rejuvenation depended almost entirely on its success in attracting new industries in large numbers. In an age when more than ever before time means money, good and fast communications were essential if industry and commerce were to be courted. Consequently the unbridged Severn estuary was a colossal barrier to economic achievement.

'There is no doubt that the Severn bridge, itself forming part of the M4 link to London, and its associated motorways has dramatically improved the situation. Growth and prosperity has throbbed back to the contracting veins of the South Wales towns and valleys as government incentives (and restrictions elsewhere) have compounded the value of the bridge.'

These statistics speak for themselves:

Over 90 000 000 vehicles have crossed the bridge since its opening. The highest number of vehicles crossing in one day is 51 641

(May 29, 1967). The highest number of vehicles crossing in one week is 278 429 (Aug 31, 1973). Over £7 500 000 has been collected in tolls to date (compare with the cost of building the Humber bridge, £60 million. The toll to cross the Humber bridge is higher than that for the Severn Bridge. Why?)

There is a maintenance team of 30 continuously engaged on upkeep, especially combatting corrosion by salt water, and a team of 38 toll collectors. Is the Severn experience likely to be repeated on the Humber?

Railway building

In 1975 the 150th anniversary of the first passenger train which

5.1 Projected major road network for Europe in 2000 AD

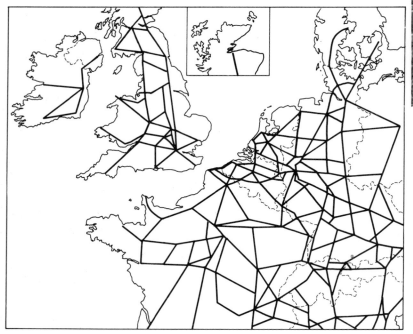

5.1a The Severn Bridge

ran from Stockton to Darlington was celebrated. In its early years the railway network grew rapidly as numerous people and companies invested their money in what was then a revolutionary form of conveyance. It reached its greatest extent about 1900 but since then competing forms of transport, particularly the car, higher costs, greater affluence of the population and many other reasons have contributed to a marked reduction in the network (fig. 5.2). The activities of the former companies were brought together as British Railways in 1947 when they were nationalised. One of the chairmen gained such a reputation for the severity of the cuts that his actions were known as the 'Beeching axe'. Further cuts

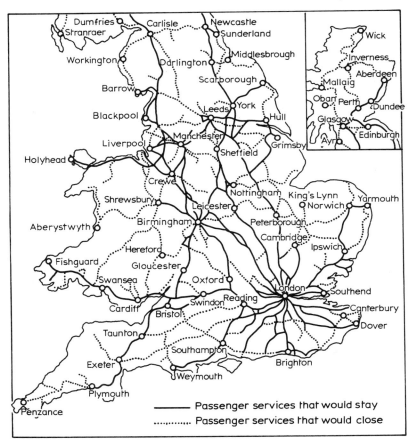

5.2 Planned British Rail network 1984

Map legend:
——— Passenger services that would stay
·········· Passenger services that would close

The advertisement in fig. 5.3 shows how they are trying to attract more business in the most profitable sectors.

1 What positive advantages are they claiming for rail transport?
2 What disadvantages are they pointing out in competing forms of transport?
3 What forms of transport are competing with the railway?
4 **Either** design an advertisement which compares the merits of railways favourably with another form of transport,
or one which compares the merit of another form of transport favourably with railways.

One of the most successful campaigns has resulted in increased inter-city passenger traffic. New stations have opened, such as Parkway near Bristol, and others outside city centres, with huge car parking provision. The map in fig. 5.4 shows the inter-city network.

are still occurring and the present network is only a skeleton compared with it at its peak.

British Rail have cut the uneconomic services and focused their attention on the most profitable operations: freight traffic, inter-city passenger traffic, and commuter traffic.

5.3 An advertisement to attract rail freight users

The Overground

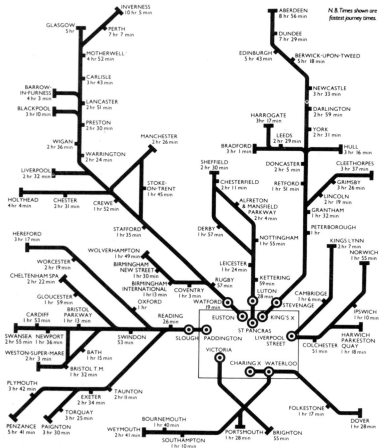

N.B. Times shown are fastest journey times.

➤➤ Inter-City makes the going easy

5.4 Inter-City rail network in Britain

The times will be reduced even further when the Advanced Passenger Train (APT) comes into general service at speeds of 200k/h.

1 What is similar between out-of-city stations such as Parkway and out-of-town hypermarkets?
2 Draw lines joining places from which it takes the same time to reach London. The distance of each from London is shown alongside the name. Are those from which it takes the same time to get to London all the same distance from London?
3 What effects are these differences likely to have on house prices, land values, demand for property, train punctuality, newspaper prices, car prices?
4 Which cities have been left out of the inter-city network? Can you suggest any reasons?
5 Describe the effects of the closure of a stretch of railway on the towns and villages along the route.

The revolution of speed

Man's speed remained unchanged until the Age of Steam. News could not travel more than 100 miles a day. No one could break this barrier. Soldiers, explorers, merchants and pilgrims moved at 20 miles per day. Paris-Toulouse required about 200 hours in Roman times, and the scheduled stage coach still took 158 hours in 1782. Only the 19th century accelerated man. By 1830 the trip

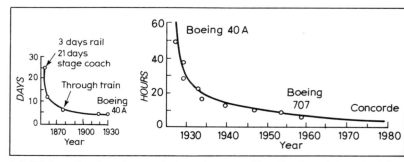

5.5 USA: reductions in travel times between New York and Los Angeles

had been reduced to 110 hours, but at a cost. In that year 4150 stage coaches overturned causing more than 1000 deaths! Then the railway brought a sudden change. By 1855 the train travelled at 96 kilometres an hour. Within a generation the Frenchman had speeded up 130 times. Today the journey from Paris to Toulouse takes 6½ hours by train.

Moving from place to place

In the mid-nineteenth century the journey from New York City to Los Angeles took 24 days. In the eighteen-seventies the travelling time had been reduced to less than a week. In the nineteen-twenties it fell sharply again. It now takes less than four hours. (fig. 5.5)

1 Consider this information and account for the speeding up of the journey over the past hundred and twenty-five years. Write a paragraph in support of your theory mentioning the time taken for the journey in 1850 and 1880, when first the journey took less than one day and the present time for the journey. Make use of the graphs in fig. 5.6 and any other information you can find.
2 What effects do you think this reduction in travel time has on people's activities?

There have been internal as well as international air movements for many years, but it is only relatively recently that they have introduced commuter services where passengers can purchase tickets in the same way as boarding a train. In North America it has been commonplace for some time because distances between leading cities are so great that there are appreciable savings in time in going by air rather than train. But in France and Britain savings are not so great because distances are much smaller and the time taken from airport to city centre reduces the difference between air and rail transport. It is forecast that Vertical Take Off and Landing aircraft (VTOL) operating from near city centres will further revolutionise inter-city travel. Until this becomes reality the demand for new and enlarged airports continues.

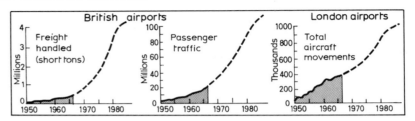

5.6 Predictions of increase in air travel

Where to put a new airport

The photograph in fig. 5.7 shows one of the newest and most modern airports in the world. Charles de Gaulle airport 24 km. north of Paris was opened in 1974. The striking architectural style of the main building has been likened to a drum on stilts, a ripe Camembert cheese and a wedding cake. Inside it is a machine for moving the passengers effortlessly to the aircraft. The traveller drives up to a kiosk, unloads the baggage onto a conveyor, collects a boarding pass, parks the car in one of 4000 spaces and then steps on a conveyor which carries passengers through a giant tube to within a few paces of the aircraft.

It is built on 3000 hectares but only 25 people had to be relocated. Some 500 in surrounding villages may have to move from the area if noise levels become intolerable. The French were fortunate to have such a large space so close to the capital to complement the other two airports, Orly and Le Bourget. The new airport

5.7 Charles de Gaulle airport, Paris

is necessary to meet the growing demand for air transport in Paris which is expected to rise to 60 million passengers by 1985.

London has a similar problem. The two existing airports at Heathrow and Gatwick are congested. As far back as 1953 it was realised that these two airports would be inadequate and steps were taken in planning an additional airport to serve London and the south east. The government repeatedly stated during the 1960s that by 1972 Heathrow and Gatwick would be choked to death. 1972 has gone, there is no third London airport, and they are still living.

Telecommunications

Look at fig. 5.8.

1 Which country has the highest number of telephones, the highest number of telephone conversations, and the most telephones per person? Why?
2 Which countries have the next highest number of telephones and make most use of them per person, on average? Why?
3 Which countries have a fairly large number of telephones but make comparatively little use of them?
4 Which countries have relatively few telephones but make fairly good use of those they have?

The number and use of telephones is a good index of the stage of development of a country and the standard of living. The USA, Canada and Sweden enjoy the highest standards of living in the world while less developed countries such as Brazil and Argentina are only just breaking into the 'telephone league'.

Fig. 5.9 shows the speed of penetration of television sets into households in the USA. Within 12 years nearly every household had access to a black and white television set. A similar pattern emerges for the spread of colour sets.

1 In which part of the USA did television begin?
2 In which parts of the USA did virtually every household possess a television set earliest?
3 Look in your atlas at a population distribution map of the USA and state true or false to each of these:
   a television stations were installed first in the smaller cities in the American mid-west and spread gradually north-east.
   b television stations were installed first in the conurbations and largest cities in the north-east and around the Great Lakes and spread south and west.

It is customary to assume that communication involves the movement of people from place to place by foot or vehicle. This meaning of communication was largely responsible for the origins and growth

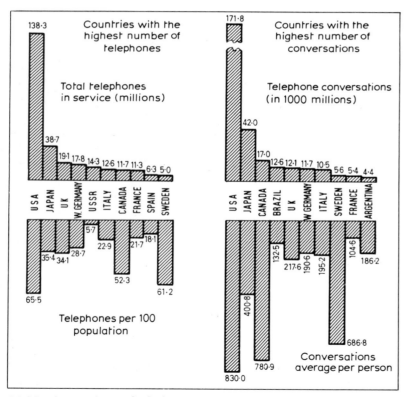

5.8 Numbers and use of telephones

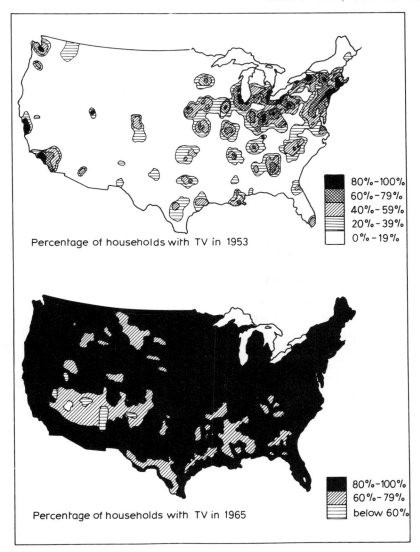

Percentage of households with TV in 1953

■	80%-100%
▨	60%-79%
▩	40%-59%
≡	20%-39%
□	0%-19%

Percentage of households with TV in 1965

■	80%-100%
▨	60%-79%
	below 60%

5.9 Market penetration by TV in USA

of settlements, but in modern times communication has assumed a new meaning. It now includes the transmission of pictures, both still and moving, and messages by television, telephone, radio, satellites and other means. Friends and relatives can keep in contact by telephone without the need to travel long distances to see each other; business transactions can be conducted by telephone, goods can be ordered from shops instead of having to visit them and so on. Already videophones are available which allow you to see the person to whom you are speaking.

The end result of these and parallel developments is likely to be a reversal of present trends, patterns and distributions. Because there will no longer be any necessity to live in urban areas people may settle permanently in the locations where they now take their holidays, choosing to live in the most pleasant rural environments which will become the most sought after locations and thus land values will rise. This process might begin in the most affluent countries such as the USA and gradually spread elsewhere. We are moving into the age of the electronic environment.

In 1902 H. G. Wells predicted that the city would spread over the countryside and assume many of its characteristics. In the same way the countryside would take on many of the qualities of the city. 'Town' and 'country' would then become terms as obsolete as stage coach.

Services

We take many things for granted because they come to us too easily: when we turn on the tap water comes out; when we put the dustbin out it is emptied; when we put the light switch down the light comes on; the roads are swept and so on. Just occasionally water does not come out of the tap, the dustbin is not emptied and rubbish piles up in the street creating health hazards and attracting vermin, candles have to be used to provide light, and the roads have remained unswept. It is only when these happen that we realise how dependent we are on these services. The high standards we experience form a major difference between the Developed and the Developing World where they are either absent or inadequate and under severe pressure. Strains are now being felt in advanced countries because:

1 The increase in population is creating a bigger demand.
2 The increase in the proportion of this population in urban areas concentrates this demand in the most vulnerable areas.
3 The physical extension of urban areas requires a corresponding extension of the services which is expensive.
4 There is difficulty in recruiting suitable labour.

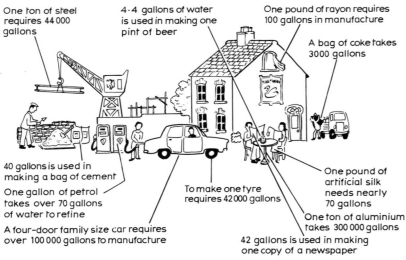

One ton of steel requires 44 000 gallons

4·4 gallons of water is used in making one pint of beer

One pound of rayon requires 100 gallons in manufacture

A bag of coke takes 3000 gallons

40 gallons is used in making a bag of cement

One gallon of petrol takes over 70 gallons of water to refine

A four-door family size car requires over 100 000 gallons to manufacture

To make one tyre requires 42 000 gallons

One pound of artificial silk needs nearly 70 gallons

One ton of aluminium takes 300 000 gallons

42 gallons is used in making one copy of a newspaper

5.10 The amount of water needed in manufacture

Water supply

Fig. 5.10 presents the problem. Read it and then answer the following:

1 How is most of the water used daily by manufacturers?
2 What makes the greatest demands on water? We drink only 0.25 per cent of the water used daily.

3 How much water is needed to produce a daily newspaper? a bag of coke?
4 What proportion of water in the diagram is used in industry?
5 Much more water will be needed in England in 30 years. List the steps that will have to be taken to ensure this supply is forthcoming.
6 List some of the features of increasing living standards that are using more water.
7 What might have to happen if we use too much water? It has been suggested that payment will be made on the basis of meter readings like gas and electricity. Belgium already has domestic as well as industrial metering.

Surplus and deficit

On the surface water appears plentiful. There are 20 million gallons available per person per day. Unfortunately only 6.6 per cent of the water that falls on the earth is used, the remainder is lost by evaporation and run off. Nevertheless in England and Wales there is still 850 gallons available for each person each day, although this must be shared amongst agricultural and industrial as well as personal uses. It is anticipated that demand will double to 28 million m³ by 2000 AD, and to meet this contingency the Water Resources Board has published a Master Plan (fig. 5.11). In the USA the average citizen is already consuming twice as much as the average Englishman. Because of increasing consumption, more and more attention is being devoted to the *management* and *conservation* of our water supplies.

Conservation is necessary to make the best use of the water we have got and to try and save some of the water that gets away. Management is needed because water has to be transferred from areas with surplus water to areas with deficient supplies. Look at fig. 5.11 which shows these areas, and the Master Plan designed to correct the imbalance.

1 Which part of England and Wales has a surplus of water?

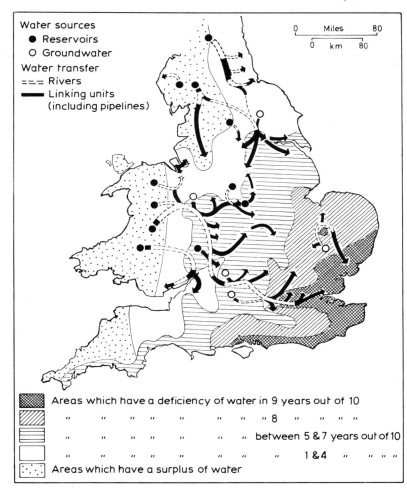

5.11 Plan for transfer of water to deficient areas, England and Wales

2 Which part has a deficiency of water?

3 Look at fig. 5.11. Use an atlas and identify London, Birmingham, Liverpool, Manchester, and West Yorkshire. These are the areas of demand. The areas of supply include

a the Elan Valley in mid Wales for Birmingham
b Lake Vyrnwy in mid Wales for Liverpool
c Thirlmere and Haweswater in the Lake District for Manchester
Measure the distance the water has to travel. How is the water transferred?

4 Why are the sources of water for these large urban areas mainly in north and mid Wales and the Lake District?

5 Why are Southern England and West Wales not included either as sources of water or as areas of demand?

6 Find examples of reservoirs which are not used exclusively for water storage. To what other uses have they been put?

Recharge

Experiments to store water underground are taking place in the Lea valley where they are artificially recharging the chalk and sand layers of the London Artesian Basin with mains water. It has to be pure because you cannot risk polluting the acquifer. Eventually it is hoped this source will produce not less than 14 million gallons a day.

Barrages

1 Draw an outline map of England and Wales with straight lines enclosing:
a the Solway Firth
b the Duddon estuary
c Morecambe Bay
d Dee estuary
e the Wash

2 The limited sites available on the mainland has led to the consideration of constructing huge dams called barrages across the mouths of estuaries and bays to enclose reservoirs. For example if the Wash barrage was built it would supply London and another city of a million people, plus furnish irrigation water for the fertile Fens. Write down the names of other estuaries which might be suitable for the developments outlined above.

3 Why have the five schemes in number 1 been forwarded first and not the others you have suggested from your atlas?

4 If the cost of the various schemes is not prohibitive which do you consider makes more sense – offshore lakes created by damming estuaries, or the damming of river valleys on the mainland in the wetter north and west of the country? State your reasons.

Waste disposal

In 1968 the average household disposed of 13 kilos of refuse a week including:

    4 kilos of paper
    1 kilo of metal
    1 kilo of glass
    $\frac{1}{4}$ kilo of plastic

In 1980 the average household disposes of 15 kilos of refuse a week including:

    6 kilos of paper
    $1\frac{1}{4}$ kilos of metal
    $1\frac{1}{4}$ kilos of glass
    1 kilo of plastic

1 How many kilos of refuse will your town produce in one week? In one year?

2 What does your town do with all this rubbish?

3 What type of waste shows the greatest increase between 1968 and 1980?

4 Why has plastic increased between 1968 and 1980?

5 What types of waste made up the difference between the four types indicated here and the total refuse disposed of by each household each week?

6 Try and keep a log of the materials that go into your dustbin at home for one week and see if it agrees with the averages above.

Waste is usually disposed of in three ways:

1 90 per cent of household refuse is tipped into innumerable holes in the ground. There are more holes than there is refuse to fill them! Refuse is deposited in shallow layers and the surface and sides are sealed with soil. Much of the refuse decays, but increasingly material is being thrown away that does not decay. Sometimes the material is pulverised before tipping. This converts it into an odourless mass unattractive to flies and vermin. Pulverisation increases the costs.

2 Composting is a fairly new and quite a costly method. A large metal cylinder (90 m. long and 12 m. wide) revolves for most of the day. All saleable material is removed by a process of sieving and rotating and items such as paper, bottles, tins, textiles and metals are sold. The remainder is placed in the cylinder and air blown in. The material decomposes by natural means.

3 There is a trend back to burning or incineration due to an increase in the quantity of paper and cardboard being used for packaging.

One of the main objectives of an efficient disposal system is to change the physical characteristics of the material so that it is no longer a danger to health and does not give rise to either private or public nuisance. Which of the above methods best fulfils this aim?

A load of rubbish

A lorry load of muck from dustbins in Nottingham smells the same as it does in any dustbin – or at least it did until recently. Now it reappears in the central heating boilers in 7000 homes and in the city centre, or rather the heat that it produces does. The transformation is illustrated in fig. 5.12. The 200 000 tonnes of rubbish burnt annually in the incinerator gives off the same amount of heat as 60 000 tonnes of coal. The hitch is that the inhabitants cannot throw away enough rubbish to meet the demand which varies from night to day and season to season. So rubbish is used as the base load and it is topped up with coal when needed. Costs are half those of conventional heating methods and the cost of collecting the rubbish is not likely to increase at the same rate as the price of coal, gas and oil. Finally a *district heating system* of this type dispenses with boilers in every home (fig. 5.12).

Although Nottingham is in the vanguard of *recycling* waste to produce heat, it is far behind many places on the continent. In the Swedish city of Vasteras, west of Stockholm, 98 per cent of the city is heated by this means. Even the pavements are heated by the one source!

Garbage into peppy piglets

Holes into which refuse has been tipped are used for a variety of purposes including housing after they have been allowed to settle. In the small south German town of Blauberen a process was developed for converting clarified sewage and refuse into humus with which abandoned quarries were filled and in which trees and bushes were planted. In this way some of the landscape of the Swabian Alps has been improved.

On the basis of what was learned a hitherto unknown method of composting was developed as illustrated in fig. 5.13. The end result is three different types of 'earth': mature compost; compost or plant earth; piglet earth.

There is a ready sale for all three products especially for piglet earth which is fed to new born piglets and chickens.

In Brittany a pilot project for the production of methane gas from fermented pig waste has been inaugurated. It appears there is a future for waste after all.

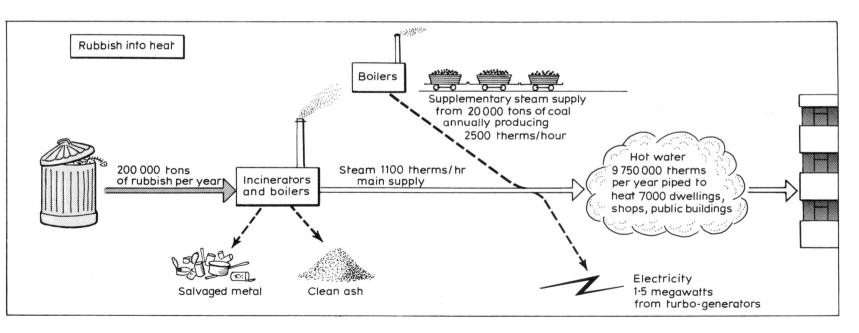

5.12 Recycling water: a district scheme

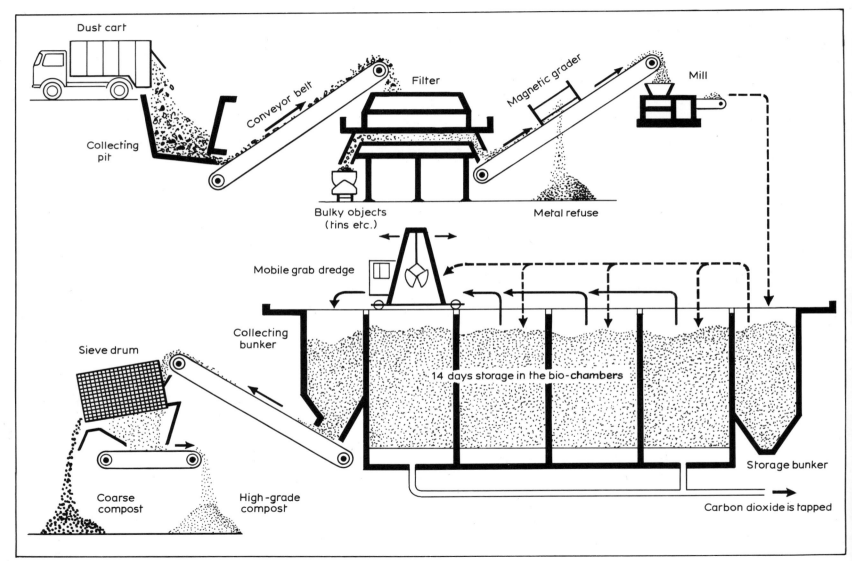

5.13 The conversion of refuse to animal food

6 Industry, Energy and Mineral Resources

Assault on the environment

The first inhabitants of Britain lived by gathering food and hunting, and their effect on the natural forest environment was negligible. The gradual conversion to sedentary farming resulted in more dramatic changes in the last 5000 years than in the previous quarter of a million years. At the time of the Neolithic farmers the density of population was less than 1 per sq km. The emergence of industry increased the density to 4 per sq km in the Iron Age and it had reached 20 per sq km by the mediaeval period. The present density of population of England and Wales of over 300 per sq km is one of the densest concentrations of people in the world. Modern farming can no longer be described as agri-culture in the sense of cultivating the soil. It is but one among many industries. Over the last 200 years the proportion of the population employed in primary extractive industries such as agriculture and mining has decreased while the proportion engaged in secondary manufacturing and tertiary service industries has increased. Now there are more people managing, servicing and marketing the product than are employed in the production of the product.

We have already considered some of the consequences of population growth and density in studying urbanisation. It remains to consider some of the consequences of industrialisation, which has promoted greater changes in the past 200 years than throughout the whole of the previous existence of man. The twentieth century consumption of power and mineral resources has been avaricious. Disasters such as Aberfan imprint the cost of 'progress' on the memory. Modern man worships the false gods in the 'temples of energy and the cathedrals of industry'. The modern wonders of the world are not natural phenomena but the spectacular man-made achievements.

Man's urge for utopia has made him more mobile. In Britain 11 per cent of the population move home each year while in the USA it is more than 25 per cent. They say they move house each time it needs a coat of paint – the ultimate in our 'Disposable Society'. Contemporary migrations in pursuit of leisure are creating a boom in tourism. All these forces cause wear and tear on a hitherto unknown scale as nature has no opportunity to recuperate. The capacity of man to utilise sensibly what are to all intents and purposes a finite land area and finite resources and to manage his environment, and the conflicts that arise between the ecologically aware and the 'technocrats', are some of the themes of this chapter.

Pollution

Company meeting focus on pollution

The International Nickel Co. of Canada Ltd. outlined efforts to reduce air pollution from its principal complex of mines and minerals at Sudbury, Ontario. Noting that the day was Earth Day, Mr Henry S. Wingate, chairman of the world's largest nickel producer, lit a cigar and devoted some time to the company's efforts to reduce sulphur dioxide pollution of the air at Sudbury and nearby Copper Cliffs. Pollution has long been an issue between the Northern Ontario community and the company. Mr Wingate reported that the company was spending 35 million dollars for 4 years to reduce emissions of the noxious gas through the construction of a 380 m high chimney stack to disperse the remaining sulphur over a wider area, about 800–1600 sq km. 'It will fall out in such minute particles as to be harmless,' Mr Wingate claimed. . . . Profits for the first quarter of the year rose to 44 million dollars.

CARBON DIOXIDE

Normally the result of energy consumed in power stations, in industry and homes. It is thought that accumulation of this gas could significantly increase the earth's surface temperature, with the possibility of geochemical and ecological disasters.

CARBON MONOXIDE

Results from incomplete fuel combustion, mostly in the steel industry, in solid waste disposal, in oil refineries and in motor vehicles. Some scientists believe this highly poisonous gas may adversely affect the stratosphere.

SULPHUR DIOXIDE

Smoke from power generating plants, industrial factories, automobiles and fuel used in the home often produces sulphuric acid. The polluted air aggravates respiratory diseases, corrodes trees and limestone buildings, as well as certain synthetic textiles and vegetation.

NITROGEN OXIDES

Produced by combustion engines, aircraft, furnaces, incinerators, excessive use of fertilizers, forest fires, industrial plants. Causes smog, may lead to respiratory infections and bronchitis in new-born children. Causes excessive growth of aquatic plants, depletion of oxygen, loss of fish and degradation of water quality.

PHOSPHATES

Found in sewage, especially in detergents, in over-fertilized land and the consequent runoff into water, and as wastes from intensive animal farming. A major factor in the degradation of lake and river water.

MERCURY

Resulting from combustion of fossil fuels, the chlor-alkali industry, electrical and paint manufacture, mining and refining processes, the pulp and paper industry, Mercury is a serious food contaminant, especially of seafood, and is a cumulative poison that affects the nervous system.

LEAD

Principal source is the anti-knock additive in petrol, but lead smelting, the chemical industry and pesticides also contribute. It is a cumulative poison that affects enzymes and impairs cell metabolism. Accumulates in marine deposits and in drinking water.

OIL

Contamination due to the operation of oil tankers, shipping accidents, refinery operation, offshore oil production and transport wastes. Has disastrous ecological effects including damage to plankton, marine life and sea birds as well as pollution of beaches and estuaries.

DDT AND OTHER PESTICIDES

Very toxic to crustaceans at extremely low concentrations. Used mostly in agriculture. The runoff of these products into the water kills off fish and their food and contaminates man's food. May have a cancer-producing effect, and may reduce population of beneficial insects, thus helping in the creation of new pests.

RADIATION

Mostly produced in nuclear fuel processing, and also in weapon production and testing and nuclear-powered ships. Has important medical and research uses, but above a certain dose can cause malignant growths and genetic changes.

6.1 The top ten major pollutants

1 Is the claim of the chairman of the world's largest nickel producer justified?

2 Does the company not have a moral responsibility to spend some of its profits on removing the objectionable emission rather than dispersing it over a wider area?

Sulphur dioxide is only one of the consequences of modern industry (fig. 6.1 lists the other major pollutants). It comes principally from burning fossil fuels which contain sulphur. Sulphurous domestic coal used to cause the London 'pea soup' fogs. In December 1952 at the height of the notorious London 'smogs' the deaths of over 4000 people in one week and over 8000 people in the next three months were directly attributable to the combination of sulphur dioxide and smoke.

1 Read these two extracts:

6.2 Some results of pollution

1968	1972
Into Bristol today comes the Prime Minister. The reason: to open a fine new plant at the Avonmouth works of the Imperial Smelting Corporation, which contains the world's biggest zinc refining complex.	The world's largest lead and zinc smelting plant—built at Avonmouth near Bristol four years ago at a cost of £14 millions—is to close for two months because of a mounting risk of lead poisoning to workers.

How can such a thing happen? Why was the risk not known before the smelter was built? Did they think people would not notice or that the problem would go away?

Water pollution – Give us our daily recycled drink

We have seen how water is a prime source of life. Fig. 6.3 shows how our water is being steadily polluted. We have seen how some of our drinking water comes from rivers (indeed it might be recycled so that the same water is drunk six times!) but some rivers such as the Trent cannot be used because of pollution. The cost of purifying the water is too expensive although it could meet the increasing demand from the whole region up to the end of the century.

In addition to the effluents which enter streams and rivers daily and poison the water, there are accidental spillages which often attract more attention than the less dramatic but more sinister week-by-week occurrences. To give an idea of the harmful effects of just a small drop of a toxic chemical imagine a football pitch under 1 metre of water. This would hold 1 million gallons. A button (or the equivalent) of the chemical endosulfan is dropped into the football pitch. This minute amount would kill all the fish in that 1 million gallons! That is what happened in the river Rhine in 1969 when millions of fish were killed. Some rivers and lakes can no longer support fish life because excess fertiliser washed into their waters have encouraged the growth of an enormous quantity of plants which have squeezed out other forms of life. Lake Erie is an example of this. The dissolved oxygen in the Baltic Sea has diminished to virtually zero and this cannot support higher forms of marine life such as fish. It is imperative that the notion that oceans are infinite sinks or cess pits for the wastes of advanced industrial societies must be eliminated before irreparable damage is done to the ecosystem.

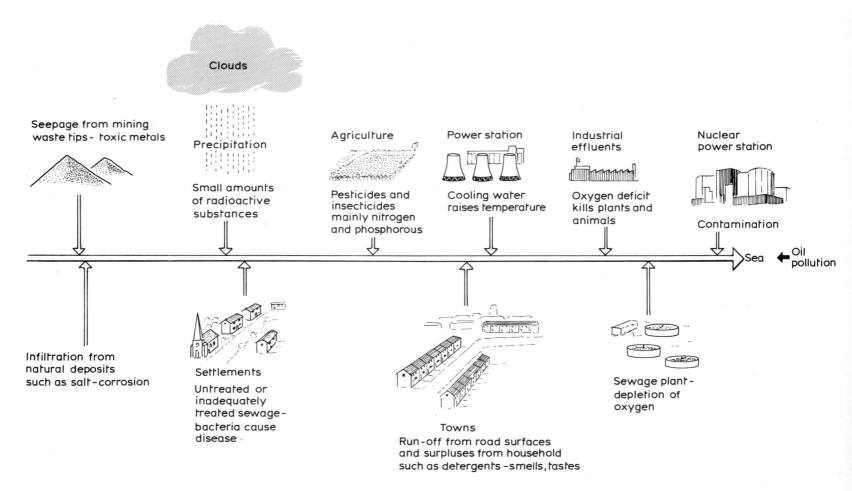

Clouds

Seepage from mining
waste tips - toxic metals

Precipitation

Agriculture

Power station

Industrial
effluents

Nuclear
power station

Small amounts
of radioactive
substances

Pesticides and
insecticides
mainly nitrogen
and phosphorous

Cooling water
raises temperature

Oxygen deficit
kills plants and
animals

Contamination

Sea   Oil
pollution

Infiltration from
natural deposits
such as salt-corrosion

Settlements

Untreated or
inadequately
treated sewage-
bacteria cause
disease

Sewage plant-
depletion of
oxygen

Towns
Run-off from road surfaces
and surpluses from household
such as detergents -smells, tastes

6.3 Some sources of water pollution

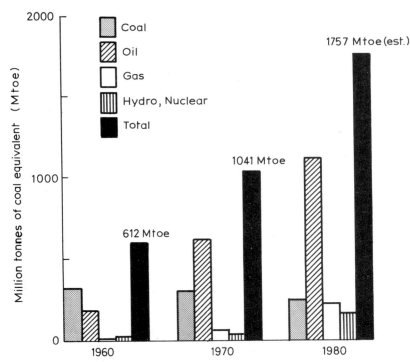

6.4 Energy consumption in Western Europe

What is being done?

The Clean Air Act 1956 virtually eliminated London smogs by making smokeless fuel mandatory and forcing industry to take stringent measures against pollution.

Although this legislation relates specifically to Britain, the problem is common to all advanced industrial nations.

Energy: Coal – a declining industry?

All industry requires energy. Water power and charcoal were supplanted by coal. Now the supremacy of coal is being challenged by other sources of energy.

Look at fig. 6.4.

1 How much energy was used in Western Europe in 1960?
2 How does this compare with the figure for 1980?
3 What proportion of the energy used in 1960 was supplied by coal?
4 What proportion of the energy used in 1980 is supplied by coal?
5 What is the major source of energy for 1980?
6 What other sources are in present use?
7 Why were these sources almost non-existent in 1960 and how have they increased their output since?

The answers to these questions summarise the pattern of energy production in Western Europe in the immediate past and the foreseeable future. On this evidence coal is a *declining* industry.

COAL MINING IN BRITAIN

Year	Tons Produced	Manpower	Collieries	Output per man-shift in tonnes
1947	185 000 000	707 500	958	1·1
1952	211 000 000	705 200	880	1·2
1957	207 000 000	707 700	822	1·3
1962	188 000 000	536 200	616	1·6
1967	164 000 000	409 700	443	1·8
1972	130 000 000	260 000	289	2·3
1977	106 700 000	242 000	239	2·4

1 Which of these statements explains the apparent contradiction that although manpower has been more than halved, and the number of collieries has been reduced by more than two-thirds, output of coal has been reduced by little more than a quarter?
a As manpower in the mines and the output of coal has declined the productivity per man shift has also declined.
b As manpower in the mines and the output of coal has declined the productivity per man shift has increased.

Why has productivity per man shift increased? It is not that the miners that remain work harder to make up for the loss of men, or that the men who have lost their jobs were not working as hard as they could, but because the power loaders which were introduced and which cut the coal and load it onto the conveyor belt operate automatically. With the help of such machines productivity is capable of exceeding 50 cwts. per man shift. The closure of uneconomic mines and those with poor seams has all contributed to the improved output per person. The contraction of an industry

in a planned way is called *rationalisation*.

As mines close in the older, worked-out parts of the coalfields others are opened in rural areas where borings and research have confirmed the existence of workable seams of coal. Figs. 6.5 and 6.6 show such an area.

6.5 Trends in the coalmining industry in Britain

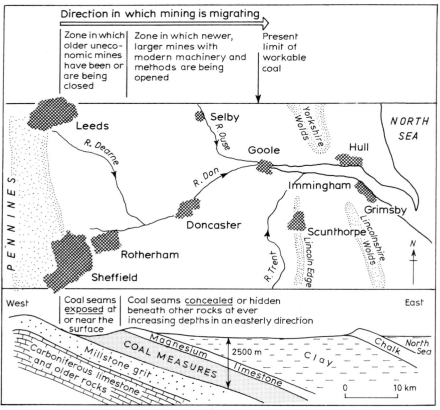

6.6 The site of a new coalmine near Selby, Yorks

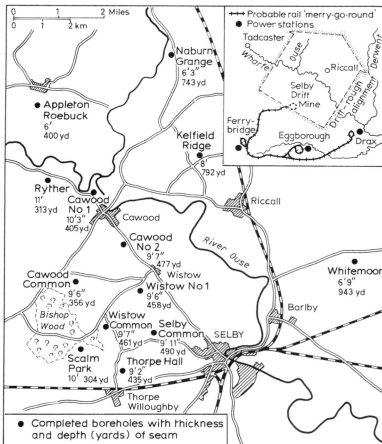

1 Suggest a site on the area shown on the map for a new mine, and give reasons for your choice.

2 Look at fig. 6.6 and state how the depth of the seam from the surface compares with the average depth of seams in England and Wales. Can you explain the difference? What has happened to the depth of mines as the years have passed?

3 How does the thickness of the seam in this area compare with the average thickness of coal seams in England and Wales?

4 What problems might confront the National Coal Board in developing a mine in the area shown on the map?

5 Why has this discovery been described as 'at least as exciting as any of the North Sea oil strikes'?

6 Why are the traders of Selby rubbing their hands and saying, 'A coalmine will be a goldmine'?

7 Why are the local councillors saying, 'These villages have stagnated for the last 20 years. We have been fighting to keep our schools open. We have poor water and electricity supplies, no gas, no shops, no bus service, no playing facilities. A coal mine certainly will not do us any harm'?

8 Why are the farmers saying, 'We will lose our livelihoods through subsidence'?

9 Why did the chairman of the National Coal Board say, 'We are fully alive to concern in the area over the environment and we intend it to be a model of environmental engineering'?

10 Use an atlas and suggest where the miners will be recruited.

Old King Coal

As we have seen, the coal mining industry has declined, although there are sufficient reserves of coal in Britain to continue production at the present rate for over 100 years. New mines are being sunk at greater depths in what were formerly known as the 'concealed' or hidden parts of the coalfield where the coal does not outcrop near the surface as in the 'exposed' part of the field. The traditional coal mining areas join the backward agricultural areas as some of the poorer regions in West Europe (fig. 6.7).

Figs. 6.8 and 6.9 indicate that the decline in the use of coal

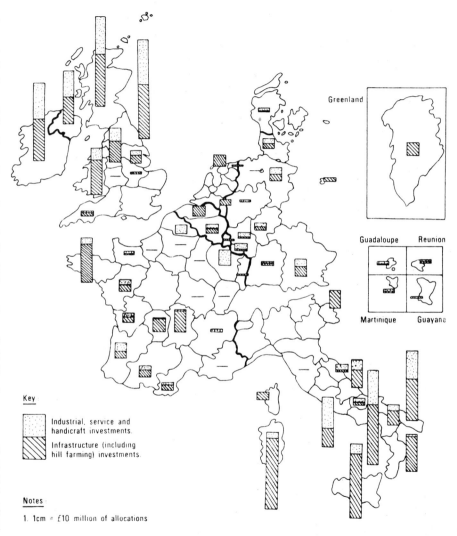

Greenland

Guadaloupe   Reunion

Martinique   Guayana

Key

Industrial, service and handicraft investments.

Infrastructure (including hill farming) investments.

Notes

1. 1cm = £10 million of allocations

Source Commission of the European Communities

6.7 Rich and poor regions in Western Europe

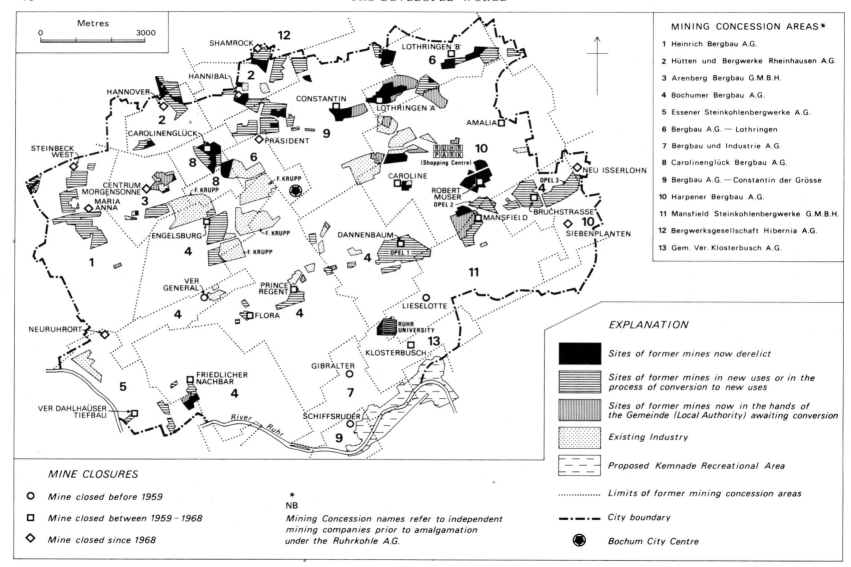

MINING CONCESSION AREAS*

1 Heinrich Bergbau A.G.

2 Hütten und Bergwerke Rheinhausen A.G.

3 Arenberg Bergbau G.M.B.H.

4 Bochumer Bergbau A.G.

5 Essener Steinkohlenbergwerke A.G.

6 Bergbau A.G. — Lothringen

7 Bergbau und Industrie A.G.

8 Carolinenglück Bergbau A.G.

9 Bergbau A.G. — Constantin der Grösse

10 Harpener Bergbau A.G.

11 Mansfield Steinkohlenbergwerke G.M.B.H.

12 Bergwerksgesellschaft Hibernia A.G.

13 Gem. Ver. Klosterbusch A.G.

MINE CLOSURES

○ Mine closed before 1959

□ Mine closed between 1959–1968

◇ Mine closed since 1968

*
NB
Mining Concession names refer to independent mining companies prior to amalgamation under the Ruhrkohle A.G.

EXPLANATION

Sites of former mines now derelict

Sites of former mines in new uses or in the process of conversion to new uses

Sites of former mines now in the hands of the Gemeinde (Local Authority) awaiting conversion

Existing Industry

Proposed Kemnade Recreational Area

Limits of former mining concession areas

City boundary

⬣ Bochum City Centre

6.8 Transformation of the mining landscape, Bochum, Ruhr

and in coal mining is common to nearly all the Developed World and is not exclusively a British problem.

1 How many pits were operating in the Borinage coalfield in 1950?
2 How many are operating there at present?
3 What has happened to the output of coal since 1950?
4 Do the remaining mines produce as much as all the mines in 1950?
5 What is the main source of employment for redundant miners?
6 How else has the area been improved to attract other industries?

A number of other steps were taken to redeploy nearly 50 000 miners made redundant since 1957.

1 Miners received full wages for 1 year while they attended special government retraining centres to teach them new skills to do an alternative job.
2 Grants for moving house and other incentives were offered for unemployed colliers to move to more prosperous regions.
3 The introduction of fast electrified train services between Mons and Brussels coupled with the traditional Belgian low-priced season ticket concession encouraged many miners to become commuters.
4 In 1959 the Borinage was declared a Development Area and industries were given incentives to locate there.

Nuclear power

Coal, natural gas and oil are *'fossil' fuels* which are not inexhaustible. They are being extracted without any replacement being returned to the ground. So, irrespective of the relative costs of the various types of energy, additional sources must be sought and utilised.

Fig. 6.4 indicates that the demand for energy is still increasing. How is this demand going to be met? France has decided to back nuclear power, but in 1980 it will only supply 6% of France's energy consumption. Britian also has a stake in nuclear power as fig. 6.10 shows, and could have 75 per cent of its electricity

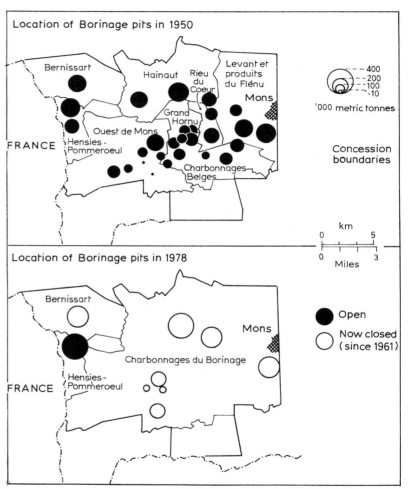

6.9 Contraction of coalmining: Borinage coalfield, Belgium

produced by nuclear power in 2000 AD, but it is hesitant to commit itself wholeheartedly to this source. There are still problems in operation, with regard to the disposal of radio-active substances and dangers of contamination through radio-activity. How do these explain the location of Britain's nuclear power stations shown in fig. 6.10?

Hydro-Electric Power

PRODUCTION OF ELECTRICITY IN WESTERN EUROPE

Country	Electricity (million KWh.)	Percentage of electricity produced by HEP
Norway	49 000	100
Sweden	49 100	94
Denmark	7 900	46
Western Germany	168 800	9
Netherlands	25 000	0
Belgium/Luxembourg	24 000	5
United Kingdom	196 000	2
Irish Republic	2 893	23
France	101 400	46
Switzerland	24 500	100

1 Which countries above derive 100 per cent of their electricity from Hydro-Electric Power (HEP)?

2 Which derive a negligible percentage from HEP?

3 In which countries are the greatest quantities of electricity produced by water? (Beware of the trap – Britain's 2 per cent is greater than Denmark's 46 per cent – statistics must be used carefully and fairly.)

4 Can you suggest why these countries have the greatest quantities or the greatest percentage produced by water? Why don't more countries make use of HEP?

Hydro-Electric Power in Scandinavia

Look at the diagram in fig. 6.11.

1 Where is the water stored?

2 What prevents it from escaping?

3 What natural feature has endowed Norway with adequate oppor-

6.10 Nuclear power stations in Britain

tunities for storage facilities like this?

4 How is the electricity generated?

5 What natural features provide the 'heads of water' necessary for the production process?

6 How is the electricity transmitted?

7 To what uses is it put?

8 Why is there less opposition to the drowning of large areas of land for storage purposes than there would be say in North Wales or the Lake District?

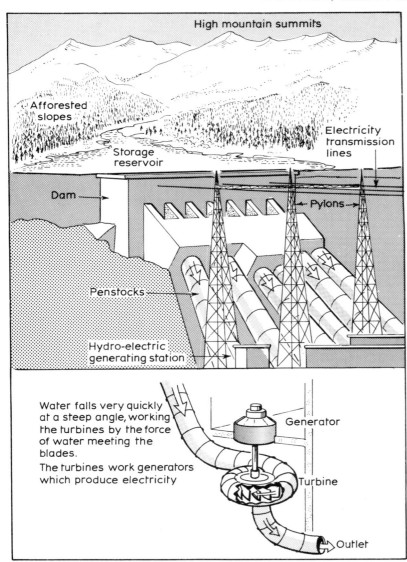

High mountain summits

Afforested slopes

Storage reservoir

Electricity transmission lines

Dam

Pylons

Penstocks

Hydro-electric generating station

Water falls very quickly at a steep angle, working the turbines by the force of water meeting the blades.
The turbines work generators which produce electricity

Generator

Turbine

Outlet

6.11 Hydro-electric power

Oil – a growth industry?

Already fig 6.4 has indicated the growing importance of energy. Production of crude oil has increased as follows:

1945	250 million tonnes
1950	500 million tonnes
1960	1000 million tonnes
1965	1500 million tonnes
1968	2000 million tonnes
1974	3000 million tonnes
1980	4000 million tonnes

Such a phenomenal rate of growth has not been achieved in any other large scale economic activity. The world's appetite for oil appears insatiable.

The primary industry of mining oil can only be located where geological conditions permit. These conditions are not equally distributed throughout the world. (fig. 6.12) Some parts of the world are thirstier for oil than others! (figs. 6.12 and 6.13) Which are they? What do they have in common? Which countries consume least oil – the Developed or Developing countries?

When oil comes out of the ground it is like thick black tar. Before it can be used as petrol for cars, fuel for central heating systems, or diesel oils for use in ships' engines and its various other uses, it has to be refined. This is carried out in a refinery where oil is heated to very high temperatures. At this stage it turns into vapour and as it is allowed to cool, it can be divided into various types of oil products such as petrol, diesel, paraffin and so on.

In the early days 50 per cent of the crude oil (mined oil) was burned or dumped because no uses could be found for it. With such a bulky product and so much waste, refineries were sited at the source so that the waste could be removed immediately, leaving only the more valuable refined oil for export. Refining oil in the early days was very much a raw material oriented industry.

With the development of cars the wasted oil had a use. In fact such was the demand for petrol that increasing quantities of crude oil had to be refined. Now they could not find a use

World oil production and consumption in million tonnes 1978

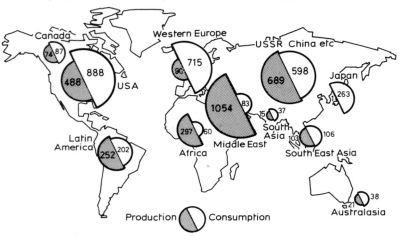

6.12 Oil production and consumption

Main oil movements by sea 1978

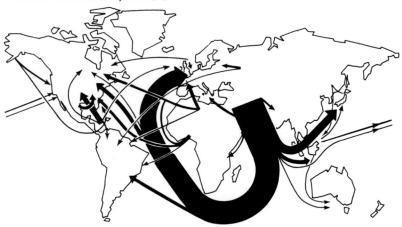

for the part of the crude oil that could not be refined into petrol! Because there was still waste, refining remained at the source of oil extraction.

Increased demand for diesel fuels gradually absorbed the waste and with the development of other uses waste was eliminated. Modern refineries have a market for virtually everything they produce. The greatest demand for oil products was in non-producing countries and as it grew the refineries near the sources of oil were unable to meet the demand. As the oil wells were owned at this time by a few large international companies, mainly American, they had a choice of where to locate the new refinery capacity needed. The new refineries were built where the greatest demands for oil existed, not where there were the greatest supplies. Oil refining had now become a market oriented industry.

Oil production and consumption

Use fig. 6.12 to answer these questions.

1 List those parts of the world which produce more oil than they consume.
2 List those which consume more oil than they produce.
3 List those which produce and consume equal or almost equal amounts.
4 Why have some parts of the world increased their capacity by much greater amounts than others?

Japan is now the world's biggest oil importer and the third largest consumer with over 263 000 million tonnes a year. Oil currently provides 55 per cent of Japan's energy requirements. It is anticipated this share will increase to 75 per cent by 1985. Yet in 1957 Japan only consumed 15 million tonnes of oil. It is trying to reduce its dependence on Middle East sources by prospecting and developing in Canada, Alaska and Indonesia and by exploration in the seas around Japan. Look at figs. 6.12 and 6.15.

1 Where is most of the oil refining capacity located in Western Europe? Why?

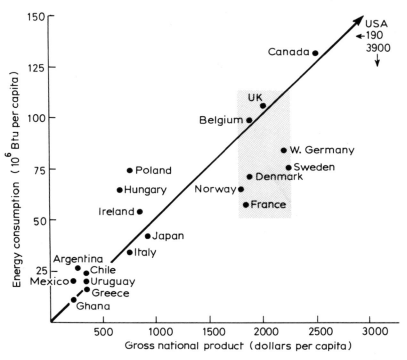

6.13 Relationship between energy consumption and gross national product

2 In how many countries is it exclusively in this location?
3 In how many countries is it exclusively in an alternative location? Why?
4 Why are there some countries with oil refining capacity in both types of location?

The transfer of oil refining capacity from the areas of supply to the areas of consumption has been accompanied by the growth in the size of oil tankers. Their size limits the ports that can accept them.

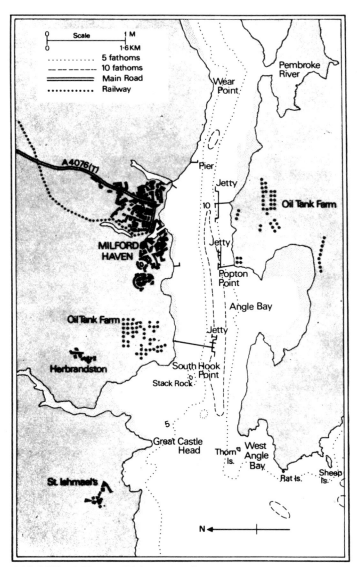

6.14 Milford Haven, Wales – an oil port

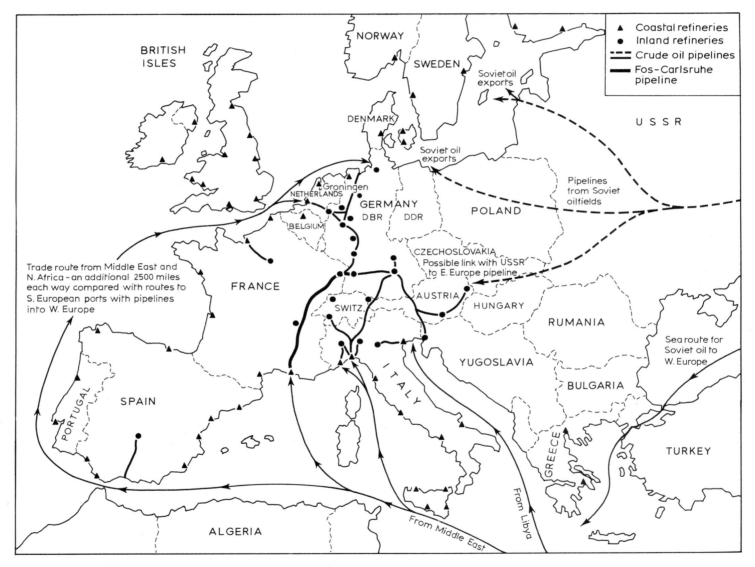

6.15 Oil pipelines in Western Europe

The location of oil refineries

The coastal location is as close to the market as the oil can get and has encouraged the location of refineries as close as possible to the point of unloading. The requirement for deep water has promoted the development of Milford Haven which is one of only 3 ports in Britain capable of handling 150 000 tonne tankers leave alone the 500 000 tonne tankers soon to be in service. (The others are Finnart in Scotland, and Immingham.) There are only 10 ports in Western Europe capable of handling these supertankers. Look at fig. 6.14.

1 What is an oil tank farm?
2 What is an oil jetty?
3 Why do the jetties stick out into the centre of the Haven? Why don't the boats berth alongside the land as in a conventional port?
4 Why is Milford Haven one of the few ports capable of taking the large oil tankers?
5 What is the name of such a natural feature and to what does this feature owe its origin?
6 Has the conversion of Milford Haven from an area of outstanding natural beauty in the Pembrokeshire Coast National Park into an oil port detracted from its scenic value? If so why? If not why not?

The construction of pipeline terminals on the Mediterranean coast at Trieste, Genoa, and Marseilles-Fos has revolutionised oil transport. It saves the tankers 7200 km of journeys to north-west Europe and ports such as Rotterdam and Wilhelmshaven and is particularly appropriate for Middle Eastern and North African suppliers. The South European Pipeline was opened in 1963 to serve the refineries in the southern Rhineland. It had already become inadequate by the late sixties. A duplicate pipeline from Fos to Strasbourg and Karlsruhe was opened in 1973. Crude oil is generally cheaper to move than oil products but the demand is so great from a centre such as Paris that it makes economic sense to convey oil products as well by this means. Fos is connected to Paris, Geneva and Grenoble by an oil products pipeline. Other pipelines can be seen on fig. 6.15 such as those from Rotterdam and Wilhelmshaven to the Ruhr emphasising the importance of the *pipeline revolution.*

6.16 The industrial agglomeration around Fawley oil refinery

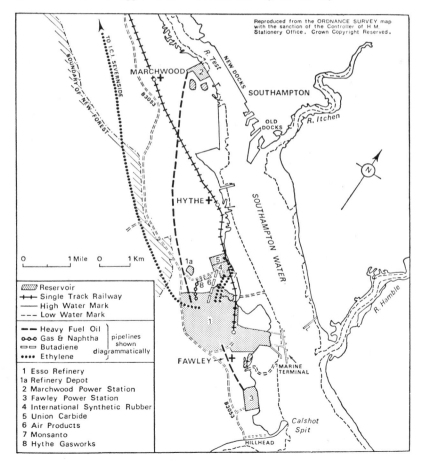

A refinery complex

An *industrial complex* is a group of technologically and economically interconnected units developed around one major industry which forms the focal part of the *agglomeration*.

Look at fig. 6.16.

1 List 3 reasons why the Esso Oil Refinery, the largest in the UK with a capacity of 19 million tonnes p.a. supplying 14% of UK market, was built in this location.
2 List 5 products produced at the refinery.
3 List 3 ways in which the oil products are moved from the refinery.
4 Complete these sentences
  a the heavy fuel oil goes by pipeline to _____ .
  b the gas and naphtha goes by _____ to _____ .
  c the butadiene goes by _____ to _____ .
  d the ethylene goes by _____ to _____ .
5 Which travels the greatest distance? How far?
6 Which is the only industry that does not receive something from the Esso refinery? This firm produces nitrogen for cleaning the vessels in the refinery.
7 Consider oil refineries in other parts of the country and abroad and see whether they have attracted other industries. Are they the same type of industries as above? Try and explain any similarities or differences which you find.

Rotterdam – Europoort

Until 1870 Rotterdam was a small fishing village. Today it is easily the world's largest port as these statistics show:

TRAFFIC IN MAJOR WORLD PORTS (in million tonnes)

Rotterdam	232	London	57	Rio de Janeiro	25
Kobe (Japan)	114	Genoa	56	Amsterdam	24
Yokohama	112	Singapore	50	Venice	23
New York	110	Hamburg	45	Gothenburg	23
Marseilles	77	Trieste	34	Wilhelmshaven	23
Antwerp	72	Vancouver	32	Bremen	22
Le Havre	61	Durban	27	Sydney	17

6.17 Rotterdam; as the number and size of ships increased, new docks were opened downstream

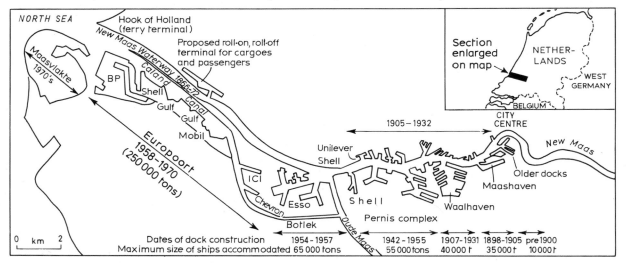

1 On an outline map of the world locate each port and draw columns proportional to the amount of cargo handled (suggested scale 2 cm = 100 million tonnes).
2 Which port handles the greatest tonnage of cargo?
3 Is there a concentration of important ports in any part of the world? Can you explain why?
4 Which of the world's leading ports are in the less developed countries? Why are there so few?
5 How has Rotterdam achieved this preeminence in 100 years?

6.18 Anyport: a model of port development

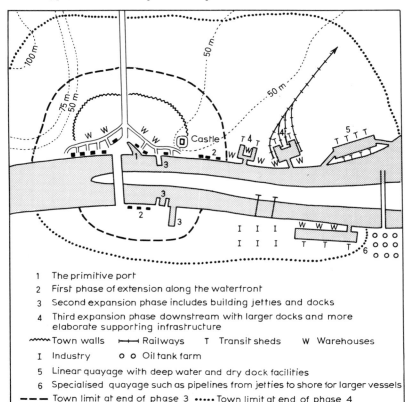

1   The primitive port
2   First phase of extension along the waterfront
3   Second expansion phase includes building jetties and docks
4   Third expansion phase downstream with larger docks and more elaborate supporting infrastructure

〜〜Town walls   ┣━┫Railways   T Transit sheds   W Warehouses

I   Industry   o o Oil tank farm
5   Linear quayage with deep water and dry dock facilities
6   Specialised quayage such as pipelines from jetties to shore for larger vessels
━ ━ ━ Town limit at end of phase 3   •••• Town limit at end of phase 4

It was initiated by:

1 The insight of civic authorities who constructed harbour facilities in advance of demand and before competing ports built similar facilities.
2 The Mannheim Convention of 1868 which allowed freedom of navigation on the Rhine previously restricted by regulations of the countries through which it ran.

The earliest 'basins' were situated close to the centre of Rotterdam and subsequent developments have proceeded downstream towards the mouth of the river where reclamation is continuing at Maasvlakte. As the newer, deeper harbours are constructed downstream to cater for more modern, larger vessels of deeper draught, the older, shallower harbours fall into disuse. (fig. 6.17)

1 What is the most important cargo handled in Rotterdam today?
2 Since when has it become most important?
3 Why has this cargo become important generally to the economies of countries and why should Rotterdam have captured so much of this trade?
4 How has the port of Rotterdam responded to this demand?
5 How is the cargo distributed to its various destinations?
6 Why does Rotterdam import more cargo than it exports?

Rotterdam's success has been due to a combination of:

1 location at the mouth of the Rhine with such a vast hinterland.
2 vigorous courtship of the multinational oil companies.
3 availability of large expanses of suitable flat land.
4 good labour relations with a record of few strikes.
5 quick and efficient handling of cargo ensuring the rapid turn round of vessels.
6 continuing reclamation at Maasvlakte at the mouth of the Rhine to provide more land for industry and port installations.

The development of Rotterdam mirrors similar processes in many ports throughout the world. It is possible to generalise the pattern of development in a 'model' which has been christened 'Anyport' (fig. 6.18).

1 Draw six rectangles in your books.
2 Using the information you have obtained, draw the phase of development reached by Rotterdam at each date.
3 Study the port nearest to you and consider to what extent it conforms to the 'Anyport' model. Account for any similarities or differences you find. To do this you will need to collect information about its origins, growth, stagnation, or decline, facilities, trade, size of vessels it can accommodate, depth of water, and other details.

North Sea Oil

Britain has placed enormous hopes in North Sea Oil. 'Nodding donkeys' scattered throughout the East Midlands constituted, until recently, Britain's oil industry producing 220 000 tonnes a year. This hardly made a dent in our consumption of 120 million tonnes a year, virtually all of which has to be imported. Then in 1964 the North Sea was partitioned and blocks auctioned to the highest bidders. Exploration of its potential began. In 1970 Sea Quest exploration rig struck oil and the 'Oil Rush' was on in earnest.

Fig. 6.19 shows the positions of the oil fields and the concessions granted for exploration.

1 How many wells had been drilled by that date?
2 How many wells had revealed their fortunes (the reserves of oil)?
3 In a similar size area off the coast of Louisiana in North America over 3000 wells have been drilled to exploit the oil. The oil companies claim that in the North Sea 300 will be sufficient, but other experts forecast at least 800 wells before the potential is fully tapped. Depending on which estimate you choose how many more wells are going to be drilled in the North Sea?

On the basis of the reserves declared, through the early 1980s Britain will produce all the oil it needs from the North Sea and have

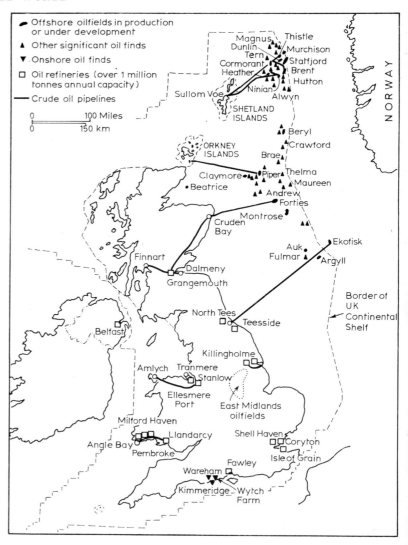

6.19 Oil fields in the North Sea

a surplus for export. Not all the financial benefits will accrue to Britain because much of the exploitation is in the hands of foreign oil firms.

These benefits have been achieved at high cost both of money and lives. When you consider the problems involved in obtaining oil from depths of over 2000 m, the equivalent of 11 or more Post Office towers, it is hardly surprising. Then there are problems of getting the oil ashore from distances of over 160 km. The costs of getting the oil ashore have escalated so much that certain firms cannot afford to exploit their proven reserves. All this makes the future prospects unclear.

New oil ports

One part of the country which is not pessimistic and is already calculating its share of the bonanza is the Shetland Islands. (fig. 6.20) Sullum Voe is certain to become the largest oil port and terminal in Britain within the next 10 years. It will handle twice the volume of traffic of Milford Haven and even outrank Rotterdam's Europoort complex. The map in fig. 6.20 shows the oil fields closest to the Shetlands which will pipe oil ashore annually. Tankers up to 500 000 tonnes will anchor in the Voe to take on either crude or refined oil.

Another major port and associated complex is to be developed on the Orkney island of Flotta on the shores of Scapa Flow to receive oil from the Piper and Claymore fields (fig. 6.19). These ports, as far from London as Corsica, will handle more than half the estimated North Sea production of 100 million tonnes a year. At this rate of exploitation it is anticipated Britain will have enough oil for 45–50 years, but it could be much longer if more wells are drilled and exploited.

The impact of oil on these sparsely populated, remote islands has already been marked. Milk rounds in the capital of the Shetlands, Lerwick, have stopped as the milkmen found jobs in the oil supply and construction bases more lucrative. A shoeshop and one of the three bakeries have closed because it was impossible to replace staff who left to work in the oil industry. A long-awaited

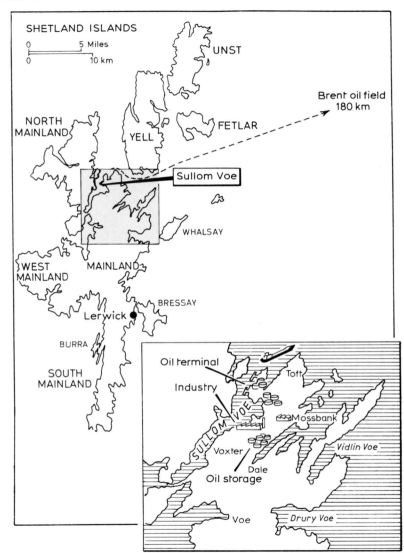

6.20 Sullum Voe, a new oil port

restaurant closed almost immediately and one of the town's two dentists shut his surgery to become a North Sea diver!

The balance of power

With the replacement of coal by oil and other fuels the expression 'the balance of power' is taking on a new meaning. Fig. 6.21 shows some of the newer sources of power in their early stages of utilisation or investigation. As yet they do not produce sufficient power between them to warrant inclusion in a country's energy output but by the year 2000 AD they could be making a substantial contribution to the energy budget as present sources become exhausted.

6.21 Alternative sources of energy

7 The Location of Industry

Many considerations have to be taken into account to discover why industries and factories are located in a particular position and place. Steel production is one such industry, and a basic one because it supplies products for other industries such as sheet steel for cars, plates for shipbuilding, girders for bridges, stainless steel for cutlery and many more. The earliest industry was located on the coalfields which were mainly in the north and west of Britain. Local coal supplied the power and the fundamental raw material, iron ore, was interbedded with the coal seams. In addition limestone, which was used as a flux, was also close at hand so it made sense to establish steel works where the main ingredients were found in close proximity to each other. Such an industry is known as a *raw material oriented industry*.

However, over the last 100 years or so there have been a number of shifts in the location of the steel industry in Britain as a consequence of many developments, summarised below:

1 Depletion of the original sources of raw materials.
2 Low quality of the early iron ore sources (about 20%+).
3 *Technological advances* which enabled the use of alternative types and sources of iron ore.
4 Opening up of new iron ore fields, particularly the *Jurassic* iron ores of the East Midlands and Lincolnshire.
5 Less reliance on domestic sources of raw materials and increasing dependence on the import of high quality ore. Fleets of huge iron ore carriers make it cheaper (per tonne) to import ore from Australia or South America than to transport it from Northamptonshire to Port Talbot. (fig. 7.1)
6 Substitution of oil for coal as the chief source of power.
7 Contraction of markets, particularly abroad as British output

has become less competitive. Over 75% of steel production is now sold in the home market. The motor car industry is the chief customer, taking 11% of total output.

Taking into account all these factors, the British Steel Corporation which manages the industry is implementing policies to *rationalise* production in fewer, larger *integrated units* instead of having coke ovens in one place, the iron works in another, blast furnaces elsewhere and the fabricating plants in yet another location. Annual production of 30 million tonnes (compare Japan 125 million tonnes a year) will be concentrated in 5 integrated units on or near the coast with access to deep water ports capable of accomodating iron ore carriers of up to 250,000 tonnes (fig. 7.2). Integration

7.1 How big ships cut costs

Cost of transport of South American ore in:

70 000 tonne lots: £2·69 per tonne

150 000 tonne lots: £1.10 per tonne

250 000 tonne lots: £1.00 per tonne

allows the industry to capitalise on the *economies of large scale production*. For example the Japanese, who have one plant capable of producing 12 million tonnes of steel annually, have an annual output of 800 tonnes per man per year compared with 250 tonnes per man per year in Britain. Their costs are also lower, being £0·15 per tonne less than in Britain. The rationalisation plan involves closure of existing works and redundancies, and the whole operation is being carried out in conjunction with the Trade Unions.

Look at fig 7.2.

7.2 Centres of steel production in Britain

1 Where are the 5 major centres of production to be located? How does the projected capacity of the largest complex compare with the largest Japanese unit?
2 Why have these been chosen? What will happen to those not chosen?
3 'Special Steel' production is to continue in an inland site where it has achieved a worldwide reputation. Where is it?
4 How many of the major centres are being developed on completely new sites known as *greenfield* sites, and how many on existing sites known as *brownfield* sites?
5 Which of the 5 major centres are in the traditional steelmaking areas and which in non-traditional areas?
6 Why are the 5 major centres of production outside the most populated part of Britain where the majority of the markets are to be found, and away from continental Europe where there are the most likely prospects of export markets?
7 If the British steel industry is uncompetitive and inefficient why bother to *invest* huge sums of money modernising production? Why not leave steel production to others and import the steel required?

7.3 A steelworks at Scunthorpe

Anchor's away

One major expansion scheme was completed and operative by January 1973 at Scunthorpe, code named Anchor, based on the existing BSC works at Appleby-Frodingham. (fig. 7.3)

1 You are the Public Relations Officer concerned with publicising and promoting and justifying the huge £235 million expansion taking place on 400 ha. of land at Scunthorpe.

Either

a prepare a publicity handout which you will send to the newspapers on the eve of the opening, **or**

b prepare an interview (or documentary) script suitable for TV or radio.

This is some of the information you will need:

The statistics are gargantuan but necessary to convey the magnitude of the operation. A crater 'lunar' landscape of old ironstone workings was levelled by 124 earthmoving machines working continuously for a year. Over 10 million cubic metres of sand and clay and $2\frac{1}{4}$ million cubic metres of slag had to be shifted – enough to bury the whole of Wembley Stadium to the top of its 40 metres high floodlights and half as high again! And that was only the beginning. The foundations involved sinking 6000 concrete piles to depths of over 30 metres. The key to the success of this massive effort are the three 300 ton Basic Oxygen converters. They are at roof top height. Another 60 metres above are the Oxygen lances to charge the converters, the anti pollution devices and other equipment dwarfing the workers. The cranes used for charging the converters with the raw materials have capacities of 450 tons and swing the ladles (106 tons) around like tea cups when they are empty. Nearly 3·5 million tons of steel will be made here in a year and all operations will be computer controlled.

Foreign ores are conveyed from Immingham 30 km away by a special train every two hours day and night. More than 100 000 tons of ore make the journey each week. The ore jetty sticks out 400 metres into the Humber estuary. The molten iron is taken from the ironmaking area to the new converters to be made into steel in huge torpedo ladle cars each filled with 250 tons of molten metal and hauled by 10 new diesel-electric locomotives. They are like giant silver cigars.

Over 1·8 million bricks were used, 16 km of new roads laid, and 48 km of new railway lines built. The conveyor belts stretch for over 21 km. The Anchor project alone uses twice as much electricity as the entire town of Scunthorpe.

Putting industry into categories

Heavy or light?

There are a number of ways of classifying industry. A simple classification is into 'heavy industry' and 'light industry'.

1 The steel industry is a heavy industry using a raw material, iron ore, which is heavy and bulky to transport, and produces a product which is usually substantial. The works cover a huge area and the whole impression is one of massiveness.

2 In contrast light industry occupies, generally, clean, one storey factories and uses raw materials which are light in weight and easy to transport. Very often they are semi-manufactured components which are put together or 'assembled' in the factory to make a finished product such as a radio, or television, or thermostat.

But there is a wide spectrum between these two extremes and it is not always easy to assign an industry to the heavy category or the light category.

Raw material oriented or market oriented

The Breisgauer Portland cement works in Western Germany uses two basic raw materials, chalk (75 per cent) and clay (22 per cent). These are crushed in the mill and made into a powder which

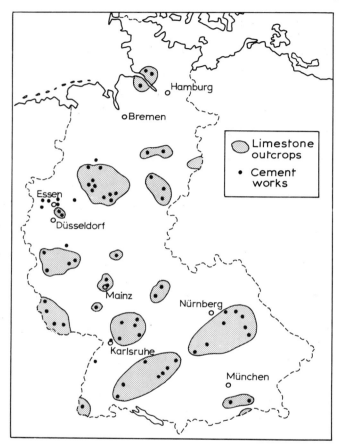

7.4 The location of steel manufacture in West Germany

is roasted in ovens at temperatures between 1400°C and 1450°C. This produces a clinker which is crushed to a powder which is cement. During the process over one-third of the weight of the raw materials is lost.

The factory uses a vast amount of power. For every tonne of cement produced some 100 KWhs of electricity and 100 Kg. of heating oil are required. The factory is fully automated and only employs 100 to operate it. A considerable amount of cement is needed in Western Germany mainly for construction purposes and in fact 680 Kg. (approx. half a tonne) is produced for every man, woman and child in the country.

1 Study fig 7.4 and copy the correct phrase. The great majority of cement works are found:
  a close to limestone outcrops
  b on limestone outcrops
  c away from limestone outcrops
  d close to other cement works
2 Copy the correct phrase. This type of industrial location is called:
  a market oriented location
  b footloose location
  c raw material location
  d geographic limit of possible production
3 Complete this paragraph by selecting the appropriate words:
  The major raw materials used in the cement industry are
  and   . It is   to transport these raw materials so
  the cement works is located   to the outcrop of
  . Since the weight of the cement is   less than the
  raw materials used it makes more economic sense to move the
  cement to markets than move the raw materials to the market and
  manufacture the cement there.
4 Look at fig. 7.5. Coca Cola is made of 90 per cent water! The remainder is sugar, carbon dioxide which gives the fizz, and the flavour which is provided by a concentrate made in Essen. The containers are disposable.
  Copy the list below into your books and
  a write 1, and 2 against the two most important factors influencing the location of the Coca Cola bottling factories.
  b put an X against the two least important factors.
  Raw materials
  Government influence
  Transport facilities
  Power supply

Market
Labour supply
Chance

5 Write a sentence explaining the distribution of Coca Cola bottling factories in this part of Western Germany.

6 Mark X and Y on a tracing placed over fig. 7.5 at two locations

7.5 Coca Cola bottling plants in North Rhine; Westphalia

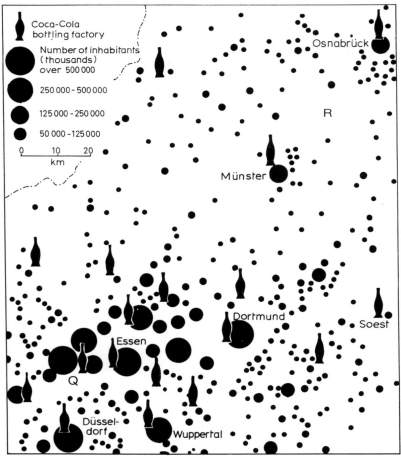

where new plants might be established. For each site explain fully why you have chosen it.

7 Explain why you would not locate a Coca Cola bottling factory at Q and R.

Putting industry in order

Another way of classifying industries is to group them according to what they produce.

Primary

Primary production includes quarrying and mining which are often called *extractive* industries. Primary production also includes agriculture, forestry and fishing because these deal directly with the natural environment. About 70 per cent of the world's population engaged in production is involved in primary production of whom 60 per cent are engaged in agriculture.

But there are important differences between countries and the percentage engaged in the various levels of production is one indication of a Developed or a Developing nation. Look at fig. 7.6.

1 Which country has the lowest percentage of the population engaged in agriculture?

2 Which has the highest?

Secondary

Secondary production involves the transformation of the primary product. Industries in this category are usually termed *manufacturing industries*. The processing of agricultural products such as canning fruit or making butter and cheese, copper smelting or steel manufacture are examples. About 15 per cent of the population engaged in production are employed in manufacturing industries. Again the percentage varies from country to country. Look at fig. 7.7.

1 Which country has the highest percentage of its population

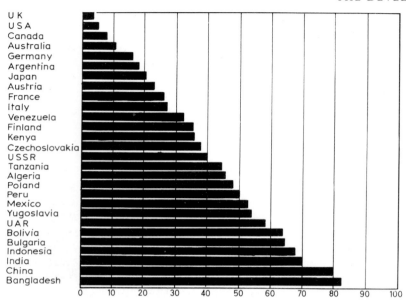

7.6 Percentage population in agriculture

and secondary establishments, and such occupations as the distribution and sale of foods and other goods. It also includes teachers, lawyers and other professionals. About 10 per cent of those engaged in production are employed in the tertiary sector although this again conceals big differences between countries.

1 Look at fig. 7.8. Which country has the highest percentage of the population employed in tertiary activities?
2 Which has the lowest percentage?
3 How many of the countries in the 'bottom 10' employers in agriculture and the 'top 10' employers in manufacturing are also in the 'top 10' of the percentage employed in tertiary activities?
4 How many of the 'top 10' employers in agriculture and the 'bottom 10' employers in manufacturing are also in the 'bottom 10' of those employed in tertiary activities?
5 Explain briefly how you can use these graphs to distinguish Developed and Developing countries.

7.7 Percentage population in manufacturing industry

employed in manufacturing industry?
2 Which has the lowest?
3 How many of the countries with the lowest percentage of their population engaged in agriculture also have the highest percentage engaged in manufacturing?
4 How many of those with the highest percentage of their populations employed in agriculture have the lowest percentage engaged in manufacturing?

Tertiary

Tertiary activity consists mainly in providing a service to products that have already passed through the primary and secondary stages. This includes the office staff in primary and secondary industries, transportation and communication workers, workers in the public utilities such as water, gas and electricity which 'service' the primary

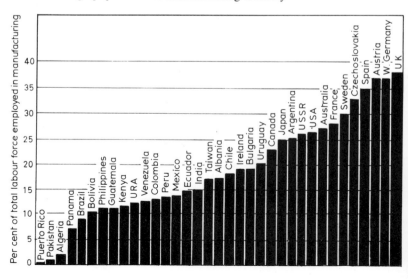

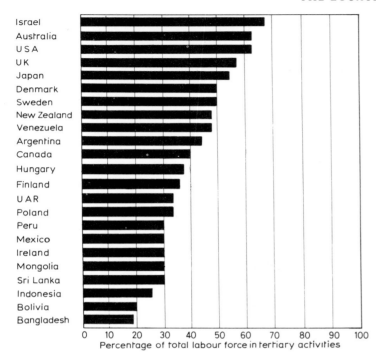

7.8 Percentage population in tertiary activities

Quaternary

This is a relatively new grouping which applies to the growing class of 'technocrats' in service occupations for which advanced training and education is required. The best known are the research establishments such as the Hydraulics Research Station at Wallingford where models of the Thames Barrage and other similar projects are tested. The importance of this sector is out of all proportion to their numbers as they are involved in the 'decision making'

end of the economic system. The 60 per cent of the world's population engaged in agriculture do not have any say in making decisions whereas the less than 1 per cent engaged in quaternary activities are actively involved. The densest cluster of innovative *high technology* industry, including silicon chips and microprocessors, is to be found in Santa Clara County, California, where over 800 pioneering companies are based around Stanford Univ. Akademgorad in the USSR is a similar scientific region.

It seems very unlikely that any classification of industry will satisfy every circumstance but they are convenient ways of handling large amounts of material and statistics.

Study fig. 7.9.

7.9 The employment structure in selected countries

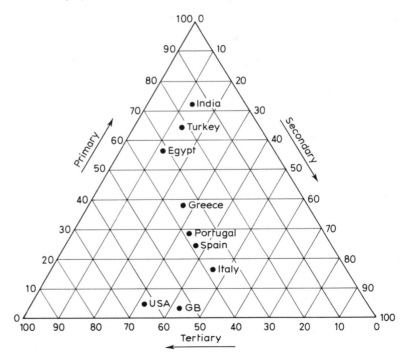

1 For each country shown on the diagram, give the percentage of the labour force engaged in each of the above categories.
2 Which countries have employment structures characteristic of Developed countries, and which of Developing countries?
3 Explain how the employment structures associated with Developing countries might gradually become similar to those in Developed countries.

7.10 The multiplier effect

One hundred new jobs in basic industry result in

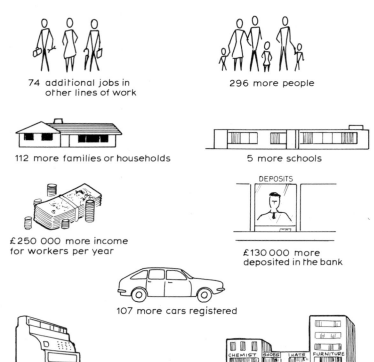

74 additional jobs in
other lines of work

296 more people

112 more families or households

5 more schools

£250 000 more income
for workers per year

DEPOSITS

£130 000 more
deposited in the bank

107 more cars registered

£170 000 more spent locally
buying all kinds of goods

4 more shops

The Multiplier Effect

Look at fig. 7.10.

1 Write a paragraph to explain what happens when a new industry opens with 100 employees. What effect does it have on the settlement in which it is located?
2 Which of these members of the local community are most likely to benefit from the opening of the factory:

bus driver
bricklayer
policeman
traffic warden
crossing control attendant
('lollypop man')
shopkeeper
undertaker
teacher
taxi driver
bank manager
estate agent
architect

3 Why have the people you listed expectations of receiving benefits from the opening of a new industry?
4 Draw a similar diagram to illustrate the likely effects of the closure of an industry employing 100 people. Give it the title 'The Reducer Effect'.
5 Which of the members of the local community listed in 2 are likely to be affected **a** most by a closure and **b** least? Why?
6 What other members not listed are likely to **a** gain, **b** lose as a result of the closure of a factory employing 100 people?

So the effects of introducing or losing an industry are not confined to the opening or closure of one building but have far reaching repercussions on the prosperity of the whole community. It is calculated that only cities with populations of over half a million

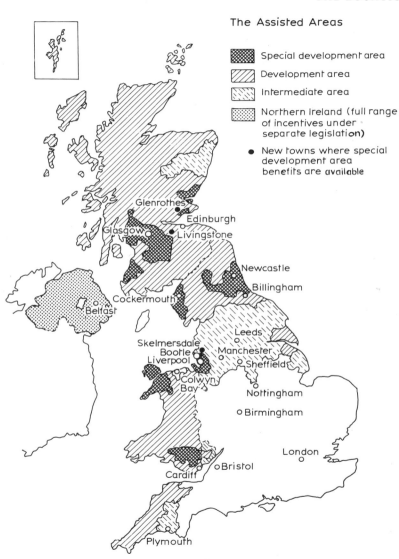

The Assisted Areas

- Special development area
- Development area
- Intermediate area
- Northern Ireland (full range of incentives under separate legislation)
- New towns where special development area benefits are available

7.11 Development and intermediate areas in Britain

can take such an occurrence in its stride without any noticeable effects. Settlements with lower populations have to take steps to remedy adverse situations such as the closure of an industry. Cities of over half a million have a built-in self-adjusting mechanism: when one industry or source of employment declines or closes, another expands and needs more workers, or a new source of employment opens up.

Development and Intermediate Areas

The map in fig. 7.11 shows the development and intermediate areas in Britain. It indicates government policy to

a get a more even distribution of different types of industry throughout the United Kingdom.

b get a more even distribution of employment throughout the United Kingdom.

Since 1948 the government has reinforced these 'persuasive' measures with deterrents and restrictions. No firm can build a factory over 280 m² in London, the south-east, or the Midlands, or over 430 m² outside these areas, without an Industrial Development Certificate (IDC) from the government. There have to be very good reasons before IDCs are granted for London and the south-east and the Midlands.

The map in fig. 7.12 shows where firms moved in response to government restrictions and deterrents to remaining or locating in high employment areas and incentives to move to high unemployment areas.

Although the government decides overall policy to locate industry in the peripheral areas, individual firms are free to choose where they locate within these areas. There is competition between development areas and between different parts of the same development area to attract industry. Fig. 7.12 shows the results.

1 Which region or conurbation lost the greatest number and % of manufacturing jobs?

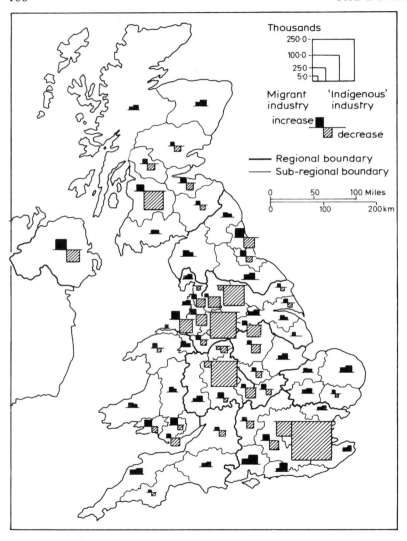

7.12 Changes in location of manufacturing industry in Britain

2 Which other conurbations and regions lost substantial numbers and % of manufacturing jobs?

3 Which parts of the United Kingdom gained the greatest number and % of manufacturing jobs?

4 Suggest some of the reasons for these changes in the distribution pattern of manufacturing employment in the United Kingdom.

5 What are the possible consequences of these trends for
a the regions and conurbations losing manufacturing jobs
b the regions and conurbations gaining manufacturing jobs.

6 What steps are being taken to arrest these trends and in some cases to reverse them?

7 An expert has commented that 'the industrial decline in Britain's cities has taken on a helter-skelter character'. Do you agree? Is it a good thing or a bad one?

Industrial estates

In 1896 the 160 wooded and landscaped hectares of Trafford Park in Manchester, for 1000 years the ancestral home of the de Trafford family, became the first industrial estate. In 1920 an industrial estate was developed in Slough. These were both private enterprises. The Economic Depression of the 1930s forced the government of the day to combat the worst evils in the badly affected areas. A distribution of industry policy dates from this period and gradually superceded the unplanned and spontaneous industrial growth which preceded it. The first manifestation of this policy was the draining of over 280 ha. of marshland to establish the Team Valley Trading Estate in Gateshead in 1936 to help solve the unemployment problem in County Durham. It was termed a 'Trading' rather than an 'Industrial' Estate because warehousing and distribution units were introduced. Today industrial estates are a common feature in many towns. Fig. 7.13 shows an industrial estate in Glenrothes.

Industrial estates are not found solely in the coalmining and traditional industrial areas. Many of them are in London and the south-east and the Midlands, and in the New Towns they often constitute the only concentration of industry. The concerns which move to the industrial estates are:

1 Industry already established in the area but looking for new premises because of expansion or redevelopment schemes in town centres.

2 Industry and commerce moving into the area from outside either voluntarily or under some compulsion.

3 Service industries catering for the day to day needs of the local population.

4 Warehousing and distribution forming part of a national or regional system.

Increasingly industrial estates are being located close to the better-class residential neighbourhoods because:

1 It reduces the journey time to work for the skilled labour and executives likely to live in the area.

2 It adds prestige and improves the company image.

Invariably the industrial estates have all the services and utilities built in so a firm only has to move and occupy the premises. In many areas estates are laid out and factories built in advance of having tenants. This is particularly true of places like Plymouth which have fluctuated between *Development* and *Intermediate* status.

Renewal of industrial properties

The restrictions imposed by the need to obtain IDCs has led to a shortage of modern one storey factories of the types found on industrial estates. This has resulted in the renovation and division of existing industrial premises. As long as each unit within the structure is below 280 m² no IDC is required so firms can locate in London, the south-east and the Midlands by taking advantage of this device.

A similar process has occurred in industrial areas in the north, but for different motives. At the beginning of the 20th century 65 per cent of the workers in Oldham were connected with cotton. Today only 12 per cent are employed in cotton, because competition from foreign manufacturers and synthetic fibres reduced the demand for natural fibres. The substantial mill buildings were ideal for renewal and subdivision into multiple uses. Mail order firms, wallpaper manufacturers, rubber producers, even battery egg production are amongst the types of industry which have located in former cotton or woollen mills (see table).

Towns which formerly relied entirely on cotton now have a diversified industrial structure and a variety of employment. It is unlikely that they will all decline at the same time. Those remaining will cushion the effects of those which close. When a decline sets in there is a 'reducer' effect which is the opposite of the 'multiplier' effect in fig. 7.10. Diversification helps to break this vicious circle.

The political factor

The policy of diversification and balance of industry and employment in Development and Intermediate Areas is a *political* decision. It is not confined to light industries. The aluminium industry is an example of heavy industry located in Development areas as a consequence of existing policy and a political decision of another kind.

In 1969 the United Kingdom produced less than 34 000 tonnes of aluminium. Present capacity is in excess of 350 000 tonnes. What has caused such a spectacular increase in capacity?

During the 1960s Britain developed a considerable aluminium fabricating business as the versatility of its products were demonstrated. It was almost entirely dependent on imported aluminium. Large import bills for the metal contributed to Britain's balance of payment difficulties. (This is like balancing an income and expenditure budget. Britain was spending too much on imports from abroad and not exporting enough to pay for them thereby constituting a deficit in our balance of payments.)

British consumption was forecast to increase to 600 000 tonnes. At this stage the government intervened and adopted an *import saving role*. We would install smelters instead of buying the smelted metal from abroad. The raw material, bauxite or alumina, would still have to be imported but this cost less than importing the smelted product.

Extract from an account of Cotton Mills closed in Bolton since 1957

Name of mill	Taken over by	New Use(s)	Labour force
Flash Street	Leslie Pink Ltd.	Occupied by 20 tenants for a variety of purposes incl. manufacture of rainwear, shop fittings, cane baskets, also for printing, storage, and offices	250–300
Bradford Mill	Demolished	Site developed as garage	25
Merton Mill	Demolished	Site redeveloped for housing	—
Derby Street	Wm. Wordsworth & Sons Ltd.	Let floor by floor — ground floor — sheet metal fabrication; first floor — furniture and bedding manufacture; second floor — engineering	100
Atlas No. 7	Courtaulds	Textiles manufacture	20
Croal Mill	Littlewoods Ltd.	Mail Order Warehouse and offices	150
Hesketh's	Asda	Superstore	300
Brownlow Fold	G.P.O.	Parcels & Sorting Office	120
Parrot Street	Cambrian Co. Ltd.	Manufacture of mineral waters	120
Garfield Mill	Bolton Corp.	Demolished for Corporation housing	—
Richd. Harwood	Demolished	Used as car park	—
North End Mill	Automotive Products Ltd.	Filters for aircraft and motor industries	500
Bee Hive Mill	Greenhalgh's	Bakery	50
Mossfield Mill	City and Country Properties Ltd.	Being let in portions for bottling and distribution of wines and spirits	20
Persian and Barley Mills	Porter, Lancastrian Co. Ltd.	Manufacture of brewery and dairy equipment, and industrial refrigeration	250
Gilnow Mill	Southern Bros. Ltd.	Manufacture of tubular steel furniture and car seats	450
Crosses and Winkworth	Elbro Ltd.	Demolished and site acquired for expansion of vehicle body building and general engineering purposes	—
Stanley Mills	Demolished	Proposed neighbourhood centre	—

The generous financial assistance to locate in Development Areas was attractive to the multinational companies in this field such as Rio Tinto Zinc, Alcan, and British Aluminium. Indeed the Lynemouth site was in a Special Development area and qualified for a 35 per cent instead of a 25 per cent construction grant. (fig. 7.14)

7.13 Industrial estate. Glenrothes. Scotland

7.14 Aluminium smelter Northumberland,

1 Why was the Lynemouth aluminium smelter located on a coastal site in a Development area? What other factors influenced its location?

2 Which expression is most appropriate to describe the location of the aluminium fabrication or fashioning plants?
a weakly market oriented, strongly raw material oriented
b weakly raw material oriented, strongly market oriented.

3 Why wasn't the aluminium smelter built closer to its users, and why is it unlikely that the users will move closer to their suppliers, the smelters?

Three base metals: one pattern

In recent years 'base metals' such as copper, tin and aluminium have joined 'precious metals' such as gold and silver as major items in world trade. The parts of the world where they are mined do not coincide with the major users of the metals. To get from one to the other they are traded between nations: the producers are usually countries in the Developing World, and the consumers are usually advanced industrial nations in the Developed World. This table confirms these generalisations:

World leading producers

Bauxite (M. tonnes)		Tin (M. tonnes)		Copper (M. tonnes)	
Australia	22205	Malaysia	64364	USA	1282
Jamaica	11304	Bolivia	28234	USSR	1100
Guinea	7620	Indonesia	25346	Chile	831
Surinam	4751	Thailand	16406	Zambia	806
USSR	4400	Brazil	5000	Canada	724
Guyana	3198	Nigeria	4652	Zaire	500
Hungary	2890	Zaire	4400	Australia	235
Greece	2850	UK	3300	Poland	230
France	2563	S. Africa	2771	Phillipines	225
Yugoslavia	2306	USA	(72)	S. Africa	178
USA	2199			Peru	173

Production

Aluminium (M. tonnes)		Tin (M. tonnes)	
USA	3519	Malaysia	83070
USSR	1500	Indonesia	17826
Japan	1016	Thailand	16630
Canada	878	UK	11585
W. Germany	677	Bolivia	7133
Norway	590	USA	6410
France	382	Brazil	5400
UK	308	Australia	5254
Netherlands	260	Spain	4700
Australia	222	Nigeria	4677
Spain	212	Belgium	4562
S. Africa	204	S. Africa	2400

Copper (M. tonnes)

Smelted		Refined	
USA	1312	USA	1620
USSR	1100	USSR	1420
Japan	821	Japan	818
Chile	724	Chile	535
Canada	496	Canada	529
Zambia	640	Zambia	619
Belgium	0	Belgium	357
Zaire	462	Zaire	304
Poland	263	Poland	248
Australia	189	Australia	192
S. Africa	178	S. Africa	86
W. Germany	168	W. Germany	422

1 On an outline map of the world draw bar graphs to illustrate the production and consumption of bauxite, aluminium, and copper. Explain the main patterns.

Figs. 7.15 and 7.16 show the *interdependence* of the Developed and the Developing World.

1 Which countries are likely to suffer most if there is a reduction in the demand for tin?
2 Which part of the world – Developed or Developing – is best able to absorb such cuts?
3 Copy the correct statements:

Tin smelting generally takes place in the same country as mining.
Tin smelting generally does not take place in the same country as tin mining.
Tin smelting takes place in countries which consume most tin. The only country which consumes large quantities of tin and smelts it is the United Kingdom.
The only producing country which does not smelt most of its mine production is Bolivia.
4 Complete this sentence using words from the list below:
Most of the tin   takes place in the   world, while most of the tin   takes place in the   world. mining Developed output capacity Developing consumption
5 In which direction is the major trade movement of tin ore or smelted tin?

7.15 International movement of iron ore by sea

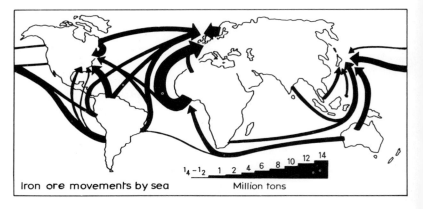

Iron ore movements by sea   Million tons

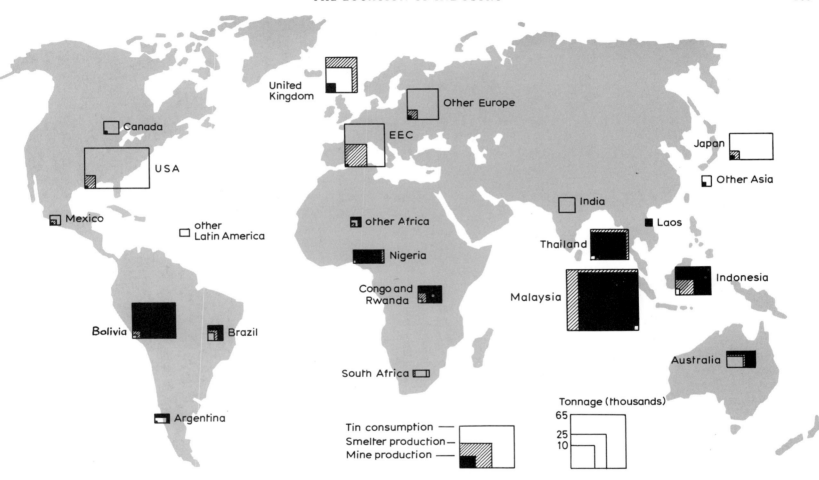

7.16 World production and consumption of tin

The major flows of these and other commodities are across the equator from south to north (fig. 7.15). Most advanced industrial consuming countries of the Developed World are in the north and the poorer primary producing countries of the Developing World are in the south. While the poorer countries have the resources which the Developed nations cannot do without, in economic and living standards they are worse off. The oil producing countries (OPEC) have shown that this can be altered.

The Producing Countries
need financial and technical assistance to develop their mines

on gaining political independence, want to lessen economic dependence

are in a position of strength in possessing the raw material, but must ensure that consuming countries do not develop substitutes if prices are raised

The Consuming Countries
usually provide the necessary financial and technical assistance

have a vested interest in protecting their investments in former dependencies

need the raw material but are willing to seek alternatives, i.e. plastic containers rather than tin cans if the price of tin goes too high

8 Leisure, Recreation and Tourism

Tourism is big business. It amounts to over 7% of world trade – exchanging people not goods! As with world trade there is a balance of payments. British tourists spend £1100 millions abroad while foreign tourists spend £2100 millions in Britain which is the equivalent of 4–5% of Britain's earnings from exports. This means Britain has a favourable balance of tourist income over expenditure. Over 11½ million tourists visit Britain each year and a similar number of Britons go abroad – but spend less! What promotes this flow of people from one country to another, and one part of a country to another?

Tourist trends

Traditionally the British have spent their annual summer holiday at one of the seaside resorts which grew up during the Victorian era over 100 years ago, when the railways connected the large populations of the industrial towns to the coast. There is a trend away from this pattern as other areas such as the Mediterranean can offer a more reliable diet of the principal ingredients of sun, sand and sea which attract people to the coast. For example Spain attracts over 30 million tourists a year who spend over £1000 million. This demand has established a southerly flow of tourists from the cooler northerly latitudes to the warmer sunnier climates. Additionally, some people now have a winter break as well as a summer holiday, and changing fashions have promoted the growth and popularity of skiing resorts in alpine regions.

Most visitors to Britain make for London because of its historic interest and international reputation. Tourism has become one of London's major industries, employing one eighth of those employed in manufacturing industries. It has also helped to replace many of the jobs in the central area which have been lost as a result of the migration of manufacturing and office employment to the suburbs. However, even in the peak season one in six beds is vacant in London because the supply of accommodation exceeds the demand for it. This is known as *overcapacity*. Besides attracting more tourists to fill these vacancies, the tourist boards are trying to persuade them to visit other parts of Britain and spread the benefits of their spending more evenly. They could make a valuable contribution to areas of high unemployment, such as the High Pennines and North Devon where tourist growth points are being developed to exploit the historic and scenic attractions, as well as rejuvenating seaside resorts.

Passing the time

How do you spend your time? For one week keep a record of the way you spend your time and compare it with the rest of your class. Are your records similar? Why?

The duration of time available for leisure pursuits is increasing as the working week gets shorter and the amount of money available for leisure purposes increases. What do people do with their leisure time?

The most popular leisure activity, television, takes place inside the home. This is a 'passive' or 'spectator' activity because it only involves sitting down and watching. Can you think of other passive or spectator activities? It is possible to 'participate' in 'active' activities in and around the home such as gardening, decorating, carpentry or a number of other hobbies. But space is limited and it is customary to go outside the home to undertake active pursuits such as games playing, riding, or swimming.

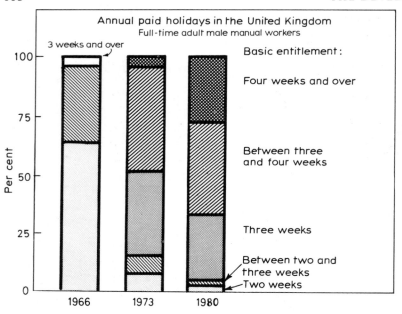

8.1 Annual holidays in Britain

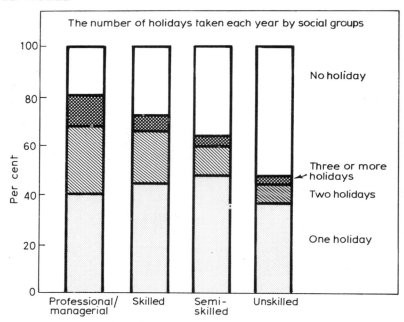

8.2 Duration of holidays in Britain

The length of holidays

National travel surveys were first conducted in 1951. This table shows what has happened since that date.

NUMBER OF HOLIDAYS TAKEN BY THE BRITISH PEOPLE
(in millions)

Year	in Britain	Abroad	Total
1951	25	1·5	26·5
1955	25	2·0	27·0
1960	31·50	3·5	35
1965	30	5·0	35
1970	34·50	5·25	40·25
1975	40	10·0	50

1 What has happened to the number of holidays taken by British people in these 25 years?
2 Which have increased by the greater percentage, holidays at home or holidays abroad?

Not only have the *number* of holidays increased but the length of each person's holiday period has grown as fig. 8.1 demonstrates.

1 What percentage of the British population had less than two weeks holiday in 1966? In 1951 the percentage was much higher and a large percentage had only 1 week.
2 What percentage had three weeks or more holiday in 1966?
3 What percentage have two weeks or less at present?
4 What percentage have three weeks or more at present?

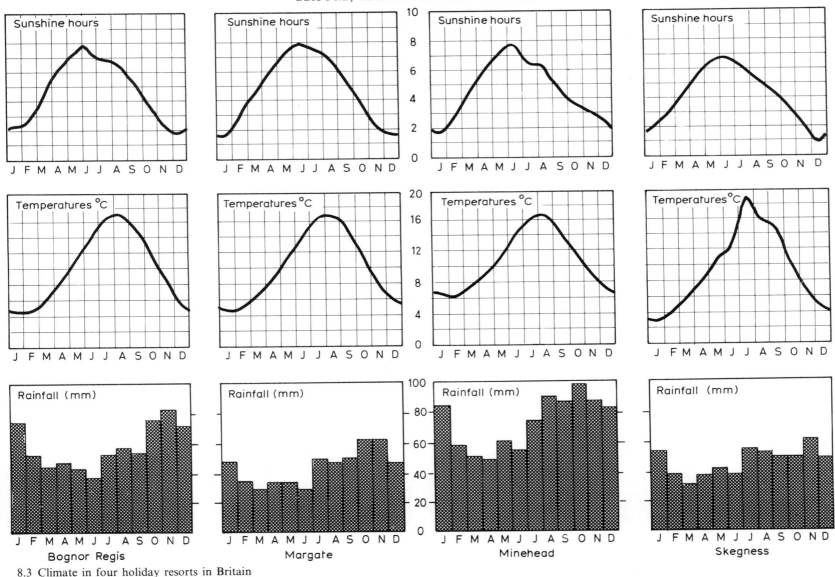

8.3 Climate in four holiday resorts in Britain

But that is not all. People used to take one holiday in their week or fortnight off work. Fig. 8.2 shows that some people now take two or even more holidays. Which social group takes the most and which the least number of holidays?

A Butlin's holiday camp

The graphs (fig. 8.3) show the climate in places where Butlin has located some of his holiday camps. Use them to answer these questions.

1 Which resort has the sunniest summer?
2 Which has the most rainfall in the year?
3 Are sun and rain important factors to the prospective client?
4 Which resort has the lowest rainfall in the main holiday season?
5 Which has the greatest temperature range (difference between highest and lowest temperatures)?
6 Which has the mildest winters?
7 Why do Butlin's seasons run only from May to September?
8 If holiday camps were open all the year and you were able to take two holidays:
　a which would you visit in summer and why?
　b which would you visit in winter and why?

The presence of such a large camp makes an impact on a settlement the size of Bognor Regis (32 000) as it does on similar size places with Butlin's holiday camps such as Filey, Pwllheli, Skegness or Minehead. Seasonal labour is recruited locally and nationally. At the end of the season some leave to find work elsewhere but those who remain are registered as unemployed.

Bognor Regis is one of many seaside resorts, which are essentially a British invention. The coming of the railway transformed them from the preserve of the genteel, well-to-do to the playgrounds of the masses. It acted as the *catalyst* to development so that they are nearly all monuments to the Victorian age. It brought the pleasures of the rich within reach of the working classes as the railways touted for trade with cheap day tickets to the coast: a day return to Skegness from London cost only the equivalent

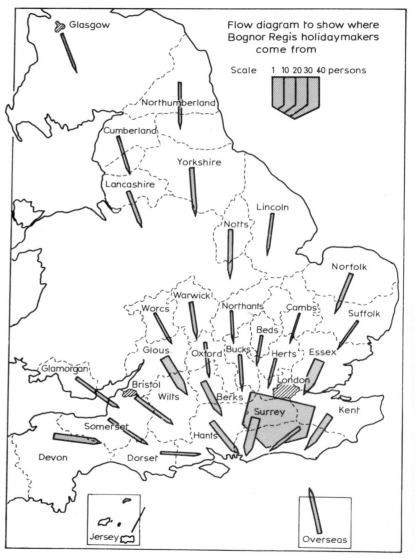

8.4 Catchment area for a seaside resort

of 15p! Eastbourne, Skegness, and Llandudno are only three of the many resorts 'planned' on formal grid iron patterns by the landowners, the Cavendish, Scarborough, and Mostyn families respectively. Their example has now been copied all over the world.

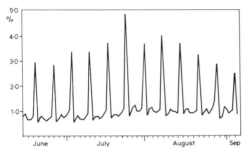

8.5 Percentage of rail passengers to Devon in summer

Origin of visitors

In the same way as towns and cities have *catchment areas* and *threshold populations* (see p. 110) for their shops and services, so do holiday resorts have a catchment area and attract more visitors from certain parts of the country than others. Fig. 8.4 shows the origin of visitors to Bognor Regis.

If you were the Amenities and Entertainment Manager of the Arun District (which includes Bognor Regis), in which of these newspapers would you advertise? Why?

South Wales Echo
Liverpool Daily Post
Birmingham Post
Bristol Evening News
Croydon Advertiser
South London Times
Newcastle Upon Tyne Evening Chronicle
Portsmouth News
The Scotsman
Sheffield Morning Telegraph
Reading Evening Post
Brighton Evening Argus

Where do they go?

HOLIDAY DESTINATIONS IN BRITAIN

| | Holidays (as %) | | | Holidays (as %) | |
	Main	Additional		Main	Additional
England			**Wales**		
South West	27	21	North	7	6
South East	15	14	South	4	3
North West	8	10	Mid	2	3
East Anglia	8	7	*Total Wales*	*12*	*11*
Yorkshire	7	8	**Scotland**		
Lake District	4	3	North	4	3
East Midlands	4	4	East Central	3	3
West Midlands	3	4	West	4	2
Greater London	3	3	South West	2	2
Northumberland/Durham	2	3	Edinburgh/Lothian	2	1
Thames Valley	2	3	Border	1	1
Total England	*77*	*78*	*Total Scotland*	*12*	*11*

1 Which is the most popular region for taking holidays?
2 Which for the second holiday?
3 Which other parts of Britain are popular for holidays, both main and additional?

When do they go?

TIMING OF HOLIDAYS IN BRITAIN

| | Main Holidays (as %) | | | Additional Holidays (as %) | |
Month	1951	1970	1975	1970	1975
May	4	5	5	16	14
June	17	6	14	12	12
July	32	33	33	8	9
August	32	32	31	19	15
September	11	10	12	19	22
Other months	4	4	4	26	28

1 Which are the most popular months for holidays?
2 Which are those most popular for additional holidays?
3 Why do most people go on holiday in two months of the year?

How do they go?

TRANSPORT USED TO REACH HOLIDAY DESTINATIONS IN BRITAIN

Means of Transport	*Main Holidays* (as %)				*Additional Holidays* (as %)		
	1951	*1965*	*1970*	*1975*	*1965*	*1970*	*1975*
Car	27	60	68	70	68	69	74
Bus/Coach	27	21	15	14	18	14	12
Train	46	19	13	13	12	13	13
Other	0	0	4	3	0	3	1

1 Which is the most popular means of reaching your holiday destination in Britain, and how has it changed over the years?
2 What is the chief means of transport for additional holidays?

The problem is that most people go to the most popular areas at the same time of the year by the same means of transport. There is a *seasonal* rhythm to the tourist industry with predictable consequences. Not only do they mostly go to Devon and Cornwall in July and August by car but once there they mostly visit the *same* parts of these counties. The problem has reached such a pitch that the authorities have had to declare certain parts of the south west peninsula 'saturation' areas.

Leisure in the countryside

Fig. 8.6 portrays the parts of Britain which are protected by legislation. It includes the 10 National Parks which have been created since 1951 to preserve extensive areas of beautiful and relatively wild countryside and provide public access to these Parks. Famous National Parks such as Yellowstone in the USA and Banff in Canada have been established for over 100 years, but the idea

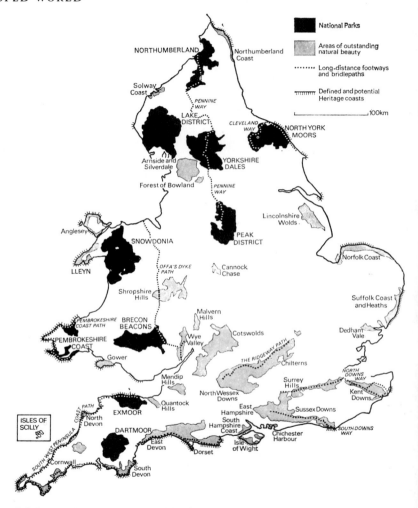

8.6 Parts of Britain protected by law

is relatively new in Britain. Unlike the USA and Canada, where there are comparatively few people compared with the land area, Britain is a densely populated country where there are many competing interests for the use of land. Amongst these is the desire to enjoy open air recreation away from the towns in which 80% of the population live. The National Parks were designated to enable their use to be planned and managed to prevent damage and deterioration of their beauty, so that this national asset could be handed down from generation to generation. However good the intention, it is not always possible to legislate for human behaviour, as this example from the South-west indicates.

To escape the crowds it seems sensible to move inland and explore the wide open spaces of the Dartmoor or Exmoor National Parks. They seem big enough for us to avoid other human beings. How wrong could we be? Far from making use of the entire extent of the Parks and walking and rambling, the majority herd together in a few popular 'tourist attractions' such as Widdicombe in the Moor. The furthest they stretch their legs is to walk around the car or go to the toilet.

The effect of the wheel and foot has been to strip the popular areas of their vegetation cover and top soil, causing erosion and visual scars. These areas which attract humans like bees are called 'honeypots' or 'gluepots'. (fig. 8.7)

1 Look at fig. 8.7 showing the intensity of recreational use in the Snowdonia National Park. Use an atlas and
   a list the 'honeypots' and 'gluepots'
   b give reasons why these are the most intensively used parts of the National Park.
2 Explain why some parts have little use or none.
3 Give examples of how the visitors can be persuaded to go elsewhere in the National Park other than the most popular areas so that the use becomes more even.
4 Do you think it is better to concentrate the majority of visitors in a few places and leave the rest relatively free, or attempt to spread them evenly throughout the Park?
5 Why are most of the towns on the edge of the National Park?
6 Using your atlas, explain why the boundary of the National Park has been drawn where it has.
7 Can you think of any reasons why there are no National Parks in Scotland?

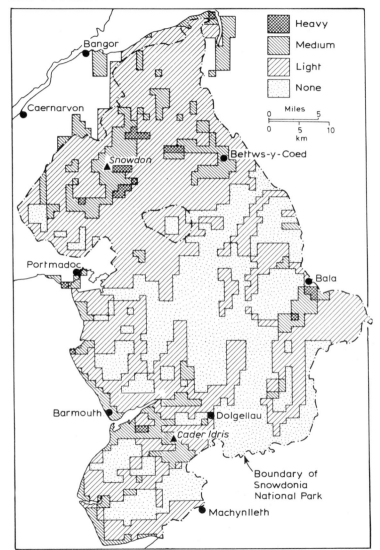

8.7 Intensity of recreational use in a national park

Positive steps, however, have and can be taken. For example:

1 The provision of carefully sited car parks.
2 The opening up of access to hill country that had been the preserve of landlords for grouse shooting. Notable successes have been Kinder Scout and Bleaklow, the two highest hills in the Peak District National Park.
3 The establishment of information centres such as the Brecon Beacons Mountain Centre outside Brecon. An anticipated clientele of 10 000 annually has exceeded 200 000 a year.
4 The creation of trails and explanatory leaflets to accompany the guides, often utilising disued railway lines such as the Buxton-Ashbourne line in the Peak.
5 Defining 'motorless zones' at peak periods such as Sundays, week ends and Bank Holidays as at Goyt Valley in the Peak. Cars must be left at the entrance to the valley and the occupants proceed by foot or in minibuses operated by the park authorities.
6 This is being extended to coaches run by the Park authorities, one coach going where it is intended being preferable to 30 cars wandering indiscriminately and adding to visual intrusion and traffic jams.
7 Establishment of residential study centres as at Losehill Hall near Castleton in the Peak District where further 'education' on the environment will take place.

Attempts are being made to manage the invasion. There is a further vital ingredient. The crowds must be 'educated' and controlled – they cannot be banned. Places like Dyfed cannot do without the £10 million annual injection which the tourists pump into its veins.

The Languedoc-Rousillon Tourist Development Scheme

The most spectacular scheme at present under way to accommodate the leisure boom with sun, sand and sea is being undertaken by the French government on the formerly mosquito-ridden marshes

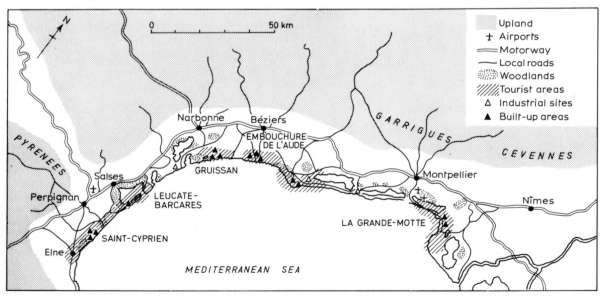

8.8 Languedoc-Rousillon tourist development scheme, France

8.9 La Grande Motte, France – a tourist resort

between the Rhone Delta and the Spanish border. Just to the south of the Spanish border the Costa Brava was the first Costa to achieve international fame or notoriety. On the Languedoc coast the French have virtually identical conditions of long uncrowded sandy beaches backed by the same Mediterranean climate and sea. They are investing £100 million over the next 20 years to accommodate 500 000 people a year.

The map in fig. 8.8 shows the road built 10 km inland connecting the river Rhone to the Costa Brava and loop roads serving a string of new tourist resorts built on the coast.

Construction is well under way and the architecture is striking and futuristic. The influence of the great French architect Le Corbusier is unmistakeable (fig. 8.9). On which British coast would it be advisable to construct such a scheme if it were ever feasible? Why?

Some influences on and of tourism

Public services

The demands of the tourists are concentrated into a relatively short period, but local authorities have to provide capacity to take the peak demands for amenities such as water supply, sewage disposal, refuse collection and so on when for the greater part of the year this capacity will be underused. On a mundane level a remote community in Central Wales had to spend 10 times more money than was necessary for their own needs on a new public convenience in the village to meet the peak demands of a short tourist season. These problems are often traced back to the seasonal nature of tourism dictated by the weather and accentuated by school and works holidays. There is a strong case for staggering the holidays and increasing the length of the tourist season. Can you think of ways this could be done, such as longer school holidays or by having four terms in a year instead of three?

Second homes

The seasonal migration of animals to summer pastures in alpine regions is a well known phenomenon. Fig. 8.10 illustrates the weekend and seasonal migration of humans to the green grass of second homes. In Great Britain nearly all of these dwellings are conversions of buildings from a variety of previous uses such as farms, barns and cottages. In Scandinavia on the other hand they are usually purpose built 'log cabins' in groups to economise on the provision of essential services. Perhaps man as well as animals has a basic instinct to return to lusher pastures as the weather improves.

Most second home owners do not travel long distances to their destinations. In Sweden they have prohibited the creation of more second homes within 1 hour's driving time of Stockholm. In Sweden 1 family in 5 has a second home. In France there are over 1 million second homes for a population about the same size as ours. The French tradition of high density apartments with little open space has encouraged the desire for 'a place in the country'. These places have become available as a result of the high rates of rural depopulation and migration to the urban areas.

It is believed that 10 000 properties a year of the type suitable as second homes become available in Britain each year. Excess demand has to be met by new construction. This demand will be particularly heavy in coastal districts, as most of the properties coming on the market are inland.

Regional flows to second homes in England and Wales

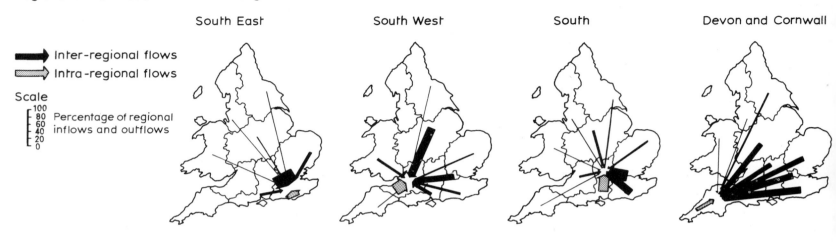

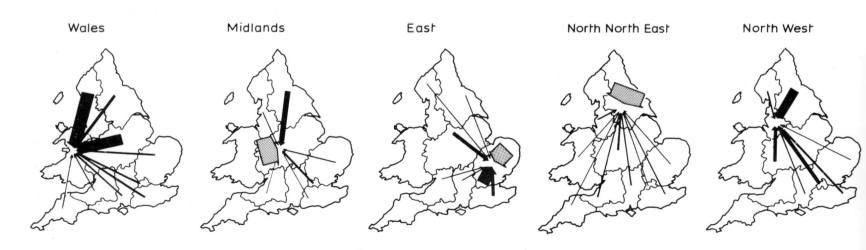

8.10 Second homes in England Wales showing regional flow

1 Briefly describe the main features of the pattern of migration to second homes in England and Wales.

2 Can you suggest reasons why the dominant flows originate in the same or immediately adjacent regions?

Leisure for the city

Country Parks

The demand in our urban areas for leisure is met by local authorities in town parks and open spaces with recreational facilities. More recently they have sponsored, along with private individuals and groups of people, a new concept – the Country Park. (fig. 8.11)

Over 100 Country Parks have been established in Britain since the government provided grants in 1968 for basic facilities such as parking, toilets and a warden service. They vary in size from Fell Foot in the Lake District which is 7 hectares to Clumber Park in Nottinghamshire which is 1521 hectares. Over 13 000 hectares is involved, of which about half was not in recreational use before. The Countryside Commission is particularly keen to encourage the conversion of derelict land, as in the Dare Valley in South Wales, formerly coal tips, and the Cotswold Water Park created from old gravel pits. No hard and fast rules are laid down as the intention is to sponsor variety. Imaginative projects have been instituted; Cheshire have reclaimed 12 km of disused railway to create the Wirral Country Park, and in Co. Durham at Hardwick Hall a derelict 18th century mansion and landscaped park is being given a face lift.

1 Where have most of the country parks been established? Why?

2 Where are they absent? Why?

3 For whom are they designed to cater?

4 If you live near a country park, visit it and list its facilities. It would also be interesting to conduct a survey to find out:

   a on which day of the week it attracted most visitors.

   b which amenities were used most frequently.

   c where the visitors originated – how far had they travelled?

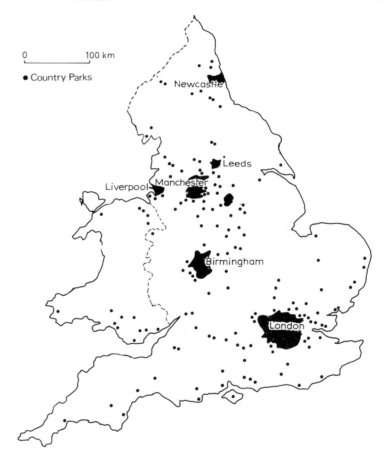

8.11 Country parks in England and Wales

Were they town dwellers? Have they, as was intended, reduced the risk of damage to the countryside?

Regional Parks – Lea Valley

The Lea Valley was the first area in Britain to be designated

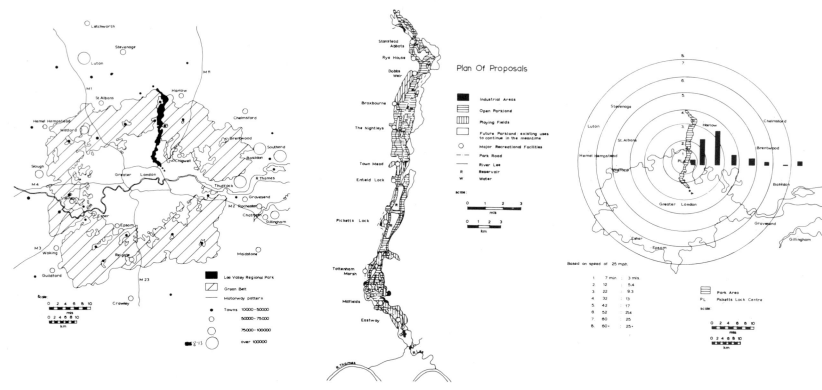

8.12a The Lee Valley Regional Park   8.12b Plan of proposals   8.12c Distances travelled to regional park

a Regional Park (1967) although the idea was envisaged in Patrick Abercrombie's Greater London Plan 1944. It is intended to transform this predominantly industrial suburban area into a vast planned leisure complex offering the widest scope of recreational, cultural and educational needs to as large a population as possible, in this case London. The catchment area of the Park is shown in fig. 8.12.

It is necessary to provide leisure facilities where the population is located. As fig. 8.12 shows, people are not prepared to travel long distances to participate.

This demand from urban populations has been met in a variety of ways, some of which are shown in fig. 8.12.

1 Choose either of these statements and write a paragraph agreeing or disagreeing with the point of view.
  a 'Planning for leisure provides an opportunity to wipe out the scars of derelict land.'
  b 'Provision for leisure is a present day bandwagon being promoted by those with vested interests and not really reflecting the desires of the population.'

9 Agriculture

A hierarchy of farming systems

The ladder in fig. 9.1 is useful because it illustrates the succession of types or systems of farming to be found in the world. It starts at the bottom with the most primitive *subsistence* systems and climbs up to the most advanced systems at the top. Although each rung marks a step up the ladder or hierarchy, the uprights of the ladder connect each step. If there were no uprights, each farming system would be separate from the one above and the one below, and it would collapse. The results of these connections and relationships are the patterns of the distribution of *land use zones* found at various levels throughout the world. The ladder is a 'model' which simplifies the complexity of the reality of the world and provides a framework to study agriculture and try and explain the patterns and distributions on the earth's surface.

The lowest rungs on the ladder are usually associated with primitive societies living at subsistence or near subsistence levels in the Third World. They are studied in the parallel book, *The Developing World*. However, until relatively recently, similar examples could be found in the Developed world.

Farming at the margin

In Welsh, 'haf' is summer and 'hafod' is the summer house. Families moved to these temporary summer houses to graze cattle and sheep on the rough mountain pasture. They also cut peat for fuel in the winter. The low land around the village was used for growing crops and fodder for winter feed for the animals. It was too valuable to be used as summer pasture. This practice of moving cattle to summer pastures was abandoned in Wales about the same time as self-subsistent farming began to change to commercial farming about 100 years ago.

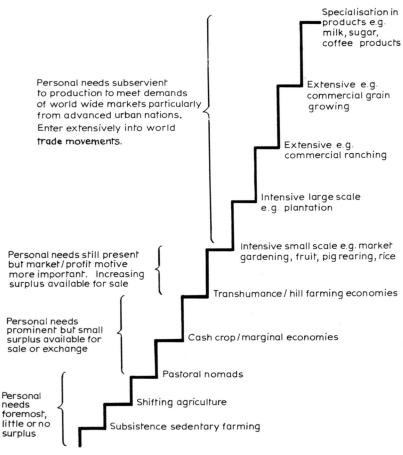

9.1 A hierarchy of farming systems

Fig. 9.2a is a 'model' devised to assist understanding of a basic pattern or rhythm of life in mountainous environments today and over a much wider area in the past. It appears that this seasonal migration to summer pastures, known as *transhumance*, is still going on. On the evidence of the Welsh experience (and elsewhere in Britain such as the Pennines and Highlands) we can formulate a hypothesis: 'Transhumance is still continuing in mountainous areas where farming is still self-subsistent.'

1 The traditional way of life

Look at fig. 9.2a.

1 At what time of the year did the cattle remain indoors?
2 Why did they?
3 When did the families begin the migration to summer pastures with their animals?
4 How did they get to the summer pastures?
5 Where did they stop on their journey to the alps?
6 Where did they halt on their return journey?
7 Why did they need to halt on their journey to and from the alps?
8 At what height was the highest summer pasture (alp)?
9 How long was it possible to graze the animals at this height?
10 What happened to the milk on the summer pasture?
11 What was growing on the fields around the village while the family and animals were on the alp?
12 What was it used for?

In this way the families grew the fodder for winter feed for the animals around the village and made cheese and butter on the summer pastures and brought it back with them for the winter. It was a self-subsistent economy.

2 The contemporary way of life

Look at fig. 9.2b.

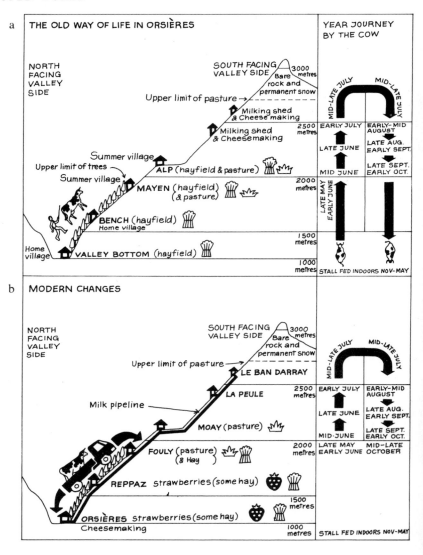

9.2 A model of transhumance, past and present

1 At what time of year do the cattle remain indoors? Is it the same as in the traditional system?

2 Why do the cattle still have to remain indoors during these months?

3 When does the migration to summer pastures begin?

4 Does the whole family go as in the traditional system? Who goes with the animals?

5 How are the animals and their handlers transported to the summer pastures?

6 Do they have to stop on the way to and from the alps? Why not?

7 What has happened to the *mayen* where they used to stop on the way up and on the return journey?

8 At what height is the highest summer pasture?

9 How long is it possible to graze the animals at this height? Why? Is it longer than in the traditional system? Why?

10 What happens to the milk on the summer pasture?

11 What is grown on the fields around the village?

12 Who looks after this crop?

13 Where does the winter feed for the animals come from now?

14 How do the people remaining in the village get fresh milk in summer if the cattle are all on the summer pastures?

15 Draw a column for the traditional way of life and one for the modern way of life and record the differences.

Until the 1930s the Val D'Entremont and Val Ferret in the Swiss Valais had been virtually self-sufficient like many other alpine valleys. Then the strawberry was introduced from the Rhone Valley and at heights above Snowdon (over 1000 m.) it was a success. It revolutionised the way of life there. To cultivate strawberries a lot of attention is needed, and while transhumance continues, many adaptations have had to be made to the traditional way of life.

A pattern that was once common to most mountainous and hilly areas has been abandoned in some, as in Wales, but continues in a modified version elsewhere, as in Switzerland.

Another feature of the traditional way of farming which has undergone a revolution in this and other valleys in Switzerland and other European countries is *landholding*. It is the same process as the 'enclosure' of the strips in the 'Open Fields' in England which took place mainly in the 18th century and gave the hedged landscape we have today. Why is it necessary to consolidate scattered parcels of land?

Consolidation is rarely undertaken by itself. It usually forms part of a wider scheme to renew the rural economy.

The old road pattern is obliterated and new straight paved roads replace the former winding tracks. Water, electricity and other services are introduced, and dairies, abattoirs and other ventures are included in the overall scheme. It is expensive and often difficult to implement because of resistance by conservative peasant farmers who are suspicious of change, especially when instituted by government officials. Today there is an additional reason. A parcel of land can be sold to tourists for chalet development at a higher price than they could earn from that parcel in their lifetime. For reasons such as these many landholdings in Western and Southern Europe are still fragmented.

Reclamation of dry land

The Swiss experience shows what it is like farming at the *margin* – in that case limited by altitude and cold. The Badajoz Plan in Spain is an example of farming at the margin of heat and lack of rainfall.

Spain is an example of a country which is making the transition from a Developing to a Developed country. For example, it still has 28 per cent of its population engaged in agriculture, but 10 years ago it was over 40 per cent. Many of those in agriculture were living in poverty because so much of the land was in private landed estates. Migration to the towns was assuming stampede proportions. To arrest or slow down this rural depopulation and at the same time to remove poverty, the Spanish government has launched schemes which will create a new human geography. We will examine one of these schemes. Look at fig. 9.3. The problem: how to transform this desert.

9.3 A desert in Spain before irrigation

9.4 The Badajoz Regional Plan

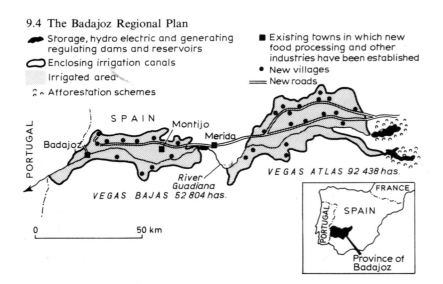

The main requirements were land and water. The land was obtained by compulsory purchase from the landlords on fair terms. The water was harnessed by investment of over £1,000 million and spectacular engineering achievements.

The Badajoz Plan

The plan was integrated, involving a number of objectives. (fig. 9.4)

1 To regulate the flow of the river Guadiana which caused havoc when in flood.
2 To transform the 'desert' of the Guadiana Plain into a fertile agricultural area.
3 To resettle peasant families on the newly reclaimed land.
4 To reafforest suitable areas especially in the catchment area of the dam to prevent soil erosion and silting.
5 To improve the communications network.
6 To establish industries in association with the new agriculture and to utilise the natural resources of the province.
7 To bring electricity and other public services to the area.

The achievements

The river Guadiana was brought under control by the construction of 5 dams which stored the surplus water and released it to the parched land when necessary through a series of canals. Before this irrigation could take place, a huge earth moving and grading operation terraced and levelled the land so that water would flow into the fields under gravity. Peasants were resettled in new villages to work the reclaimed land. Once water was supplied, the former desert burst into bloom with a whole range of fruit and especially pears. In addition cotton, peppers, tomatoes and corn are grown and animals kept for meat and milk on irrigated pastures. Industries have been established to process the agricultural produce which amounts to over $1\frac{1}{4}$ million tonnes a year. They include pear canning, tomato juice, pepper dressing, a cotton mill, packing depots, a slaughterhouse and creamery. There is also a wood distilling plant associated with the afforestation programme, and mining of the natural resources of the area.

Amenities and infrastructure

Despite comparable incomes it is doubtful if the Plan would succeed in keeping peasants on the land unless it provided services, facilities and amenities which the colonists would take for granted if they lived in the urban areas. Therefore a crucial part of the strategy has been the provision of electricity, telephones, television, doctors and dentists as well as improved roads and rail services – what is termed the *infrastructure*. In addition, social amenities and entertainment facilities have been introduced to offer a viable alternative to migration to the cities.

Inter-basin water transfers

Progress is impressive, but even more spectacular schemes are planned. In Spain it is also proposed to reverse the present north-east to south-west flow of the river Tagus and by a series of diversions and tunnels to irrigate the dry lands of Almeria and Murcia in the south-east corner of Spain. A massive scheme of this type has already been completed in the Snowy mountains in eastern Australia, where the plentiful supply of water from the eastern slopes has been channelled by tunnels and reversing the flow of rivers to irrigate the interior Murray-Darling basin. In North America there are even more ambitious proposals to transfer enough water to supply 337 cities the size of London for one year from the Arctic to the drier southerly latitudes (fig. 9.5). The need is proven because so much water is extracted from the river Colorado for agricultural and urban purposes that virtually none reaches the sea. However, the cost is enormous and alternatives such as *desalination*, weather modification, *groundwater* exploitation, *watershed management*, water rationing and pricing, are all being considered to conserve a vital resource.

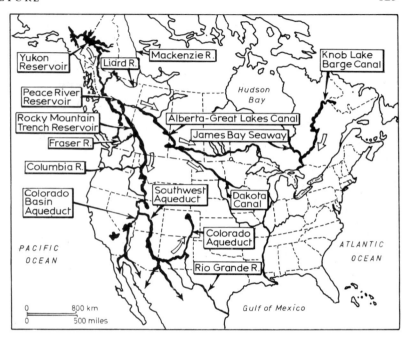

9.5 Plans for inter-basin water transfers in North America

Italian land reform

Other Mediterranean countries are in the process of transforming their landscapes through schemes of land reform. Perhaps the best known is Cassa del Mezzogiorno in Italy (fig. 9.6). Most of the investment has been devoted to southern Italy where the problems are greatest. It bears a close resemblance to the Spanish experience. Spasmodic attempts had been made before the second World War but the main effort dates from 1950. It was launched to arrest the drift of population from the south to the industrial centres of the north. It involved the confiscation of the empty lands of the great landed estates and the colonisation of this land with new settlements of formerly landless peasants. They were provided with homes, land, assistance, and the essentials to begin farming.

a   Campania (Volturno, Garigliano, Sele)

b   Apulia, Lucania, Molise

...... Boundary of Southern Italy as defined by the Cassa del Mezzogiorno

9.6 Italian land reform regions

Altogether about 8½ million ha. are involved, comprising 28% of the total area of Italy. Elsewhere, in long settled areas, consolidation of scattered parcels of land and holdings and other improvements have increased incomes and raised standards of living. As in Spain,

it is as much a social revolution to combat poverty as an agricultural revolution.

Reclamation from the sea

In Spain the problem was to get water onto the land, while in the Netherlands the problem is to remove water from the land. Nearly two-thirds of the country is below sea level. At the same time it is one of the most densely populated countries in the world. How then has the Netherlands become such an affluent nation? We have already seen that it has the world's leading port and a vibrant conurbation. Another part of the explanation is revealed in fig. 9.7.

1 When did the Zuider Zee project begin? When is it scheduled to finish? Why did they decide to reclaim land here?

2 When did the Delta project begin? When is it scheduled to finish?

3 Why is the Delta project scheduled to be completed in a shorter time than the Zuider Zee project?

4 From which city is the population likely to come to utilise the recreational facilities created by the Delta Plan?

5 Which city is likely to solve some of its housing and overcrowding problems by spilling over onto the Sud Flevoland *polder*?

6 What is the main purpose of the creation of polders from the Yssel Meer?

7 How much land will have been reclaimed from the sea by the time the Zuider Zee project has been completed? This represents one-fifteenth of the entire land area of the Netherlands.

8 Which 'New Town' is planned to be the central place of the reclaimed polders? Why has it been chosen? Why has it been given this name?

9 What principles have the Dutch planners employed in the establishment of patterns of settlement on the polders?

10 Describe the location of most of the 15 000 ha. which will be reclaimed by the Delta project.

These schemes have other advantages:

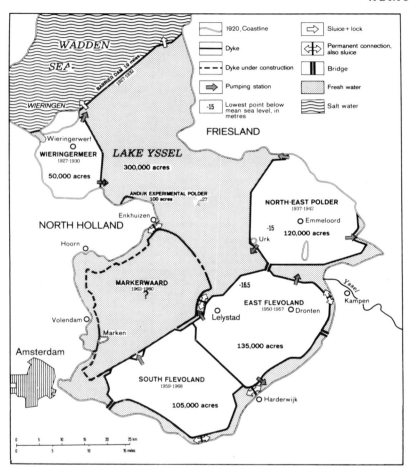

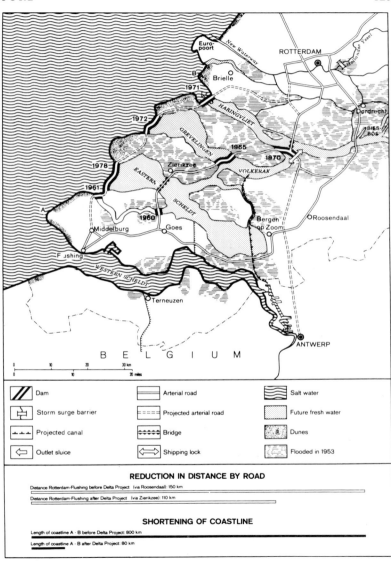

9.7 Land reclamation in the Netherlands. Above, the polders of the Yssel Meer: right, the Delta plan

1 They make the task of looking after the coastline much easier.
2 They counter salinisation.
3 They relieve the isolation of certain areas, particularly the province of Zeeland in the South-west.
4 They largely remove the risk and fear of storm floods similar to those which devastated large areas in 1953.
5 They provide fresh water lakes for recreation.
6 They provide fishing grounds, although at the expense of the disappearance of oyster culture and the ancient shrimp and mussel fishing industries.
7 Shipping channels and other communication links including roads of motorway standard and bridges are connected to the national systems.

As in Spain the reclamation of farming land is not the sole purpose. In the Netherlands the acquisition of new fertile soil appears to be a by-product of a much larger integrated multi-purpose scheme. But the provision of additional land for agriculture is vital to feed this densely populated, highly industrialised nation.

Part time farming

The temptation to give up farming and opt for an easier way of life is great. Look at fig. 9.8.

These present the choices facing farmers. Can they afford to grow crops when there are less arduous ways of making a living? In cases where the profit per hectare is not very high and the farmers do not have many hectares, it is far more lucrative to use land for camping sites and bed and breakfast places than to grow crops. The fact that these circumstances coincide with the most popular tourist areas in the north and west of Britain highlights the issue.

A survey conducted amongst farmers in North Wales found most of them reluctant or openly hostile to accommodating visitors on their farm. Their reasons were:

Lack of time, space or money for alterations or developments; tenancy restrictions, age or ill health; inadequate roads; fear

9.8 The farmer's choice. Above, growing crops: below, using the land for caravan or camp sites

of vandalism; reluctance to change the Welsh way of life, or to allow visitors to interfere with farm activities or spoil the countryside.

And yet there are sound economic reasons for participating in tourism. Hill farms, which are in the majority in many of the tourist areas, only supply 8 per cent of the agricultural output of the country and less than 1 per cent of the value of agricultural output. They take 20 per cent of all government subsidies yet their average income is still very low. It is estimated that 65 per cent of their income is accounted for by subsidies of one sort or another. Can they afford not to take tourists? Write a paragraph saying to what extent you agree with one farmer who replied: 'If farmers cannot make a living out of farming they should get out and leave it to those who can.'

Some farmers have adopted a 'half way house' by becoming part-time farmers. Over half the farm holdings in Britain are classified as 'part-time holdings'. While they are usually the smaller units, they nevertheless comprise a significant proportion of farmers. A similar trend can be detected in most Western European countries. Fig. 9.9 illustrates this trend in a village near the university, administrative and industrial city of Göttingen in the Leine valley in Western Germany.

1 How many full-time farmers were there in Hetjershausen in 1960?
2 How many were there in 1960 who had a full-time job during the day and worked their holding as a 'hobby'?
3 How many were full-time farmers in 1975?
4 How many farmers were there in 1975 who had a full-time job during the day and worked their holding as a 'hobby' in the evenings and at weekends?
5 What has happened to the total numbers of farmers between 1960 and 1975? Can you give reasons?
6 What has happened to the number of houses in Hetjershausen between 1960 and 1975? What is likely to have attracted people to move to Hetjershausen? Where would they have come from?
7 Why is part-time farming considered a 'solution' to the problem? What are the advantages and disadvantages from the point of view of a country's economy? What are the advantages and disadvantages from the point of view of the part-time farmer?

Commercial pressures and consumer preferences

People in glasshouses

Glasshouse crops are an example of a branch of agriculture affected by tourism and the associated urban expansion in a different way. It is the opposite end of the spectrum from the hill farm. It is *intensive* and not *extensive*. Glasshouse crops occupy less than 1 per cent of the land area devoted to agriculture in Britain but are responsible for more than 10 per cent of crop value. Traditionally Hertfordshire and Essex have had the largest concentration of glasshouses in the country.

A large number of the growers (why do they call themselves growers and not farmers? Have you noticed the weather forecast on the television is for farmers and growers?) have had to move from the Lea Valley and have chosen West Sussex. Why did they choose to move there rather than to other parts of the country?

The map in fig. 9.10 gives some information on the factors which have to be taken into account.

1 Rewrite the correct versions of the following:
  a The majority of glasshouses are found within 0–2 km., 2–4 km., 4–6 km., 6–8 km., 8–10 km. of the coast.
  b The majority of glasshouses are found on the 'Hook' or 'Hamble' soils.
2 The graphs (fig. 8.3) show the temperature, rainfall and duration of sunshine in West Sussex and in the other important glasshouse crops area outside the Lea Valley.
  a Which area has the highest temperatures?
  b Which has the lowest rainfall? Why is that not so important when the crops are grown in glasshouses?
  c Which area has the longest hours of sunshine?
3 Complete the following:
The majority of glasshouses are found . . . .

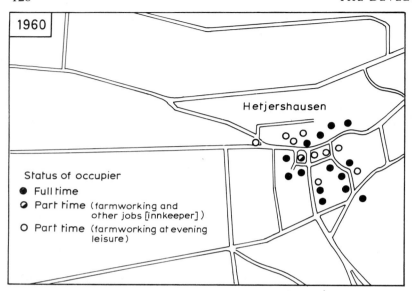

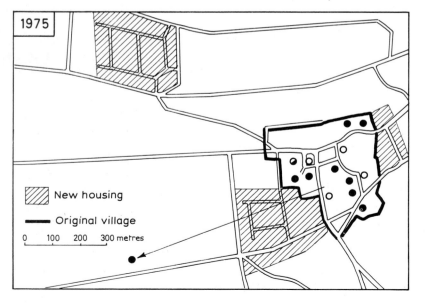

These answers indicate that physical factors such as soil and climate play a part in the location of glasshouses in West Sussex. What other physical factors have to be taken into account? Look at fig. 9.10 and remember this part of West Sussex is called the 'Coastal Plain'.

Fig. 9.10 also shows the situation of other smaller, less important but nevertheless significant concentrations of glasshouses. In terms of industrial location these would mark the *geographic limit of possible production* of glasshouse crops. Then why are some areas such as West Sussex growing faster and at the expense of others within these geographic limits?

You have already found out some of the answers but there are other reasons. Planning permission has to be obtained before new glasshouses can be erected.

In July 1966 the British government introduced a special grant of nearly 40 per cent of the cost of the erection of new glasshouses. It made West Sussex a 'Development area for glasshouses'. But this grant was available anywhere in the country, so although it explains the explosion in applications it does not explain why so many chose West Sussex.

The cost of erecting a modern glasshouse is very high (£80 000). Although he has a grant, the grower's chief concern is to get his money back as soon as possible because until he does he is not making any profit. This influences his choice of crop because he will grow the crop which gives him the greatest profit most quickly.

The crops grown in the earliest glasshouses about 100 years ago were grapes, peaches, figs and melons for wealthy clients. Gradually tomatoes and lettuce became more important to meet the demand from the growing urban population especially during the two World Wars. While they are still the most important vegetables they have been overtaken by 'luxury' crops such as flowers. Nurseries used to grow a wide range of products but increasingly they specialise in one crop. This degree of specialisation has dangers as well as benefits. What are they?

9.9 Part-time farming: adjustment near a large city

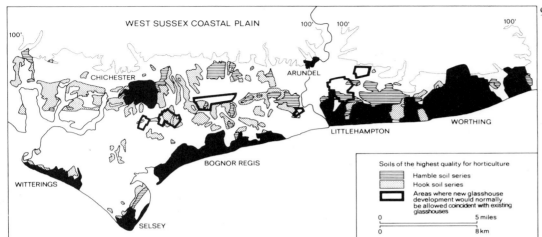

9.10 Location of glasshouses in West Sussex, England

1 Which is the more important item grown under glass, vegetables or flowers?

2 Which is the most important vegetable?

3 Which are the most important flowers?

4 How many growers grow only one item?

5 Is specialisation greatest in vegetables or flowers?

The trend is towards growing more 'luxury' crops such as flowers instead of the day-to-day needs of the population because they produce higher profits. If your father grows flowers you will know that he can only produce them once in the year and success will be dictated by the weather. But the glasshouse growers have introduced a new phrase into the vocabulary – all-the-year-round chrysanthemums or carnations (A.Y.R.). How can they produce flowers throughout the year?

Glass alone would not achieve this miracle. But these are no ordinary glasshouses – they are controlled environmental machines. Computers keep the temperature at the precise level required, allow each individual plant its own drink of water through a hole in a pipe alongside, draw the blinds to 'kid' the plants that it is dark, or turn on the lights so that they think there is continuous daylight. The grower knows when he plants each crop the precise day it will be harvested, so he can get 5 or 6 crops of flowers from one plant in two years, $2\frac{1}{2}$ or 3 crops each year! In any one glasshouse you will find crops in all the stages of the 16 week cycle it takes to grow the flower. As in the steel industry *technology* has revolutionised *productivity* in glasshouses.

But heating and lighting cost money, so growers locate where these costs will be lowest – that is where they need to use least. Most heating and lighting is needed in the winter and records show that West Sussex has the most intensive natural light conditions and the highest winter temperatures in the country. Soils are less important factors than they used to be because most crops are now grown in artificial, so called 'hydroponic', soils, used over and over again. After each crop they are sterilised by blowing steam through them.

They still have to sell the crops. Traditionally the crops were sold through Covent Garden (or Nine Elms) and distributed from there throughout the country, the biggest market being London. But there are changes here as well as in production.

9.11 A modern wide-span glasshouse

1 Some growers formed a *cooperative* which collected the produce from each individual and then sold it altogether. This saved each grower marketing his own crops and enabled larger quantities to be sold. A big organisation can bargain for prices better than one person, and can buy plants and fertilisers in bulk which is cheaper than small orders. This is an example of *economies of scale*. A cooperative is beneficial to the small producer.

2 Some of the largest firms do not need to belong to a cooperative because their organisation is large enough to enjoy the economies of scale such as obtaining discounts for bulk purchase and bargaining for good prices for their produce. They have gone further. In industry certain giant corporations have emerged which control all the stages of production from mining the raw material to selling the finished product (Rio Tinto Zinc is an example), and some large growing firms have organised themselves in a similar way. The control of all operations from start to finish is known as *vertical integration* and is illustrated in fig. 9.12.

In the case of the grower, the final product is *perishable* and must be sold close to the market. But production has expanded

9.12 Vertical integration in glasshouse crops production

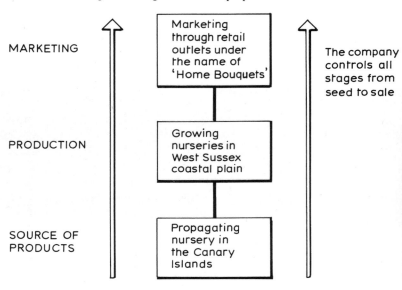

MARKETING — Marketing through retail outlets under the name of 'Home Bouquets'

The company controls all stages from seed to sale

PRODUCTION — Growing nurseries in West Sussex coastal plain

SOURCE OF PRODUCTS — Propagating nursery in the Canary Islands

to the Canary Islands, Portugal and Malta where the climate is ideal and no artificial heating or lighting is required. Workers are also plentiful and their wages are lower than in Britain. These assets help to offset the cost of transporting the seedlings by refrigerated air or sea transport to Gatwick airport or Shoreham harbour. Why to these transport terminals? Were there large home markets in these foreign countries it is likely that the entire operation would move and cease production in Britain. But there are not. So firms locate in the *optimum location* in Britain where the physical factors are most favourable, with proximity to the market. The main market is the largest centre of population – London – and they also have direct sales to shops or merchants in other large cities. But more important are their sales to their own shops (Home Bouquet) which form the end of the chain of vertical integration which began in the Canary Islands, Malta or Portugal.

Glasshouse crops are an example of an agricultural system which has rationalised production and organised itself to be competitive in regional and international markets, and would be near the top of the ladder (fig. 9.1).

1 Which of these phrases best describes vertical integration?
  a Putting together two grates in an upright position.
  b A number of firms getting together to effect economies.
  c One firm controlling all stages of production and distribution of their products.
2 Which of these phrases best describes cooperation?
  a A number of growers getting together to effect economies in purchasing and marketing.
  b A number getting together to control all stages of production and marketing of their crops.
  c A number helping each other out by sharing their machinery.
3 Which of these phrases best describes optimum location?
  a The highest location.
  b The worst location.
  c The best possible location.
4 Which of these phrases best describes the geographic limit of possible production?

a The area within which it is impossible to produce a good or crop because geographical conditions are unfavourable.
b The area within which geographical conditions permit the production of a good or crop.
c The area within which it is not permitted to grow a given product or crop.
5 Which of these phrases best describes perishable product?
  a One that has a short life.
  b One that is highly desirable.
  c One that has a long life.
6 Which of these phrases best describes productivity?
  a Achieves greater output through more effort.
  b Achieves greater output through less effort.
  c Achieves greater output through the application of the latest technical developments.
7 Imagine that you are the Planning Officer for West Sussex who has to make the decisions on the application for extensions to existing glasshouses or the construction of new glasshouses. You have decided to group the area into three 'regions': (fig. 9.10)
  a Where you would not allow additional glasshouses to be built.
  b Where you would allow additional glasshouses to be built.
  c Where you would allow additional glasshouses to be built in exceptional circumstances.
  Write a letter to an applicant in each of the categories explaining
  a why you have not granted the application.
  b why you have granted it.
  c the exceptional circumstances which led you to either accept or reject the application.

Consumer preferences – frozen food

1 Write in your own words the trends which fig. 9.13 indicates.
2 Which of the following explain the trends?
  People tiring of fresh food
  Fresh food becoming too expensive
  More wives going out to work
  People having less time to prepare fresh foods
  People liking frozen foods

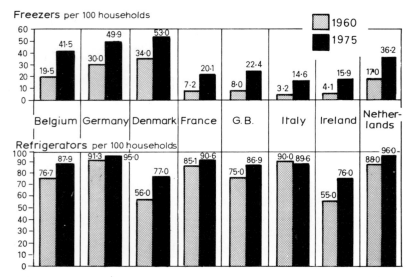

9.13 Home freezer ownership in G.B. and Europe

Ease and convenience of frozen foods
More people living alone
Husbands objecting to opening tins
More homes with fridges and deep freezes
Increasing proportion of non-active people.

When Clarence Birdseye opened his first freezing factory in Lowes-
toft in 1931 he started something. Increasing affluence made refriger-
ators status symbols in the 1960s and deep freezes in the 1970s.
This, associated with the greater numbers of working wives and
an increase in the pace of living in which time became a valuable
commodity, generated the frozen food revolution depicted in fig.
9.13. The most popular single frozen food is peas.

The maps in fig. 9.14 show where 85 per cent of the peas for
freezing are grown in Britain. We will investigate why there is
such a marked concentration in this area.

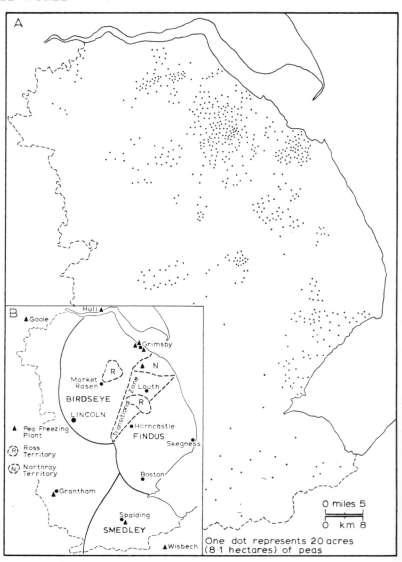

9.14 Growing for freezing: Lincolnshire

1 What has happened to the area devoted to growing peas for freezing over the past 10 years, and why?

2 Near which town is the greatest concentration of peas?

3 Look in your atlas and see if there is any marked concentration of pea growing on
  a the high or low ground?
  b any particular geological formations?

4 Where are the pea freezing plants located? Does that offer an explanation for the concentration of pea growing?

Peas are harvested over a short 8 week period between early July and late August. During that time the pea vining 'circus' migrates over the area shown on fig. 9.14. Each mobile viner harvests 40 to 50 hectares in the season compared with 18 hectares for each static viner which they replaced in 1965. This has increased both the range (13 to 32 km.) and the quantity of peas that can be grown and harvested. Fig. 9.14 shows there has been greater intensification in areas nearer the freezing plants as well as an extension of the limits of production. When a crop spreads over an area the process is called *diffusion*.

Peas cannot be grown at great distances from the freezing plants because to avoid bacteriological deterioration the time gap between vining and freezing must not exceed 90 minutes. To ensure a smooth continuous flow into the plants during the season, field operations are controlled by a mobile centre equipped with short wave radio. As soon as the peas reach the factory they are chilled. Following further processing they are put in cold storage until they are required for repacking.

The frozen pea industry is tightly linked to that of frozen fish because the peas were originally grown to take up slack capacity in fish freezing plants. The large area of cold storage available in association with the fish industry made Grimsby an admirable base for the frozen foods.

To maintain quality there must be a high degree of cooperation between farmer and processor. The farmer really 'hires' his land out to the freezing firms in return for a guaranteed price. *Contract farming* offers a security to the farmer which speculative farming does not. Previously the farmer was dependent on the vicissitudes of market prices; now he can plan ahead. Rye for Ryvita, blackcurrants for Ribena, are just two of many crops 'tied' to a processor. See if you can find others if you live near a farm.

1 Explain how technology has affected the growing of peas for freezing.

2 What controls the geographic limit of possible production of pea growing?

3 What is meant by the perishability of the pea crop?

4 Why does Grimsby offer an optimum location for the freezing of peas?

5 What is meant by diffusion?

6 What is meant by contract farming?

Political influence – Drinka Pinta Milka Day

This slogan was part of the promotion campaign launched by the Milk Marketing Board in the 1960s to get people to drink more milk. At that time Britain's farmers were producing too much milk. We are still producing the same quantity of milk but greater efficiency, such as the introduction of milking parlours instead of cattle stalls, means it is being produced by fewer farmers. There will be less than 50 000 farmers producing milk in the 1980s, which is 30 000 less than there were in the 1970s.

The Milk Marketing Board was established in 1933 and was the first of a number of government sponsored marketing bodies. It pioneered 'contract' farming by offering the farmer a guaranteed price for his milk. Furthermore it organised the collection and distribution of the milk and processed any surplus into butter, cheese, evaporated milk or milk powder. The uncertainty of fluctuations in market prices was replaced by the security of the monthly cheque.

Fig. 9.15 shows the location of the manufacturing plants belonging to the Milk Marketing Board. Compare their locations with fig. 9.14. Why are they situated in the west and north of the country?

The study of the glasshouse crops, peas for freezing and the Milk Marketing Board are all illustrations of the way agriculture

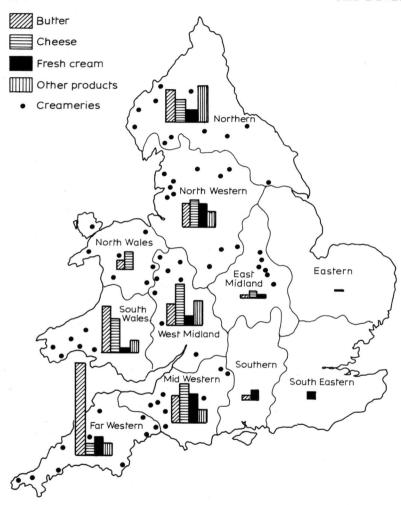

Butter
Cheese
Fresh cream
Other products
• Creameries

Northern
North Western
North Wales
South Wales
West Midland
East Midland
Eastern
Southern
Mid Western
South Eastern
Far Western

9.15 Manufacturing plants belonging to the Milk Marketing Board, UK

has become *specialised* and as a consequence has *migrated* and *contracted* towards the *optimum* locations. See fig. 9.16.

1 Copy an outline of England and Wales and on it draw a line separating those counties where barley occupies the largest area from those where rotation grasses occupy the largest areas.
2 Suggest reasons why this division occurs where it does.
3 In which counties is no wheat grown? Why?
4 Why do the counties around London have comparatively little land under crops?
5 Why do the Welsh counties have comparatively little land under crops? Is it for the same reason as around London?

The divisions you have drawn and the answers show that certain parts of the country specialise in growing certain crops. This specialisation is not exclusive because other parts of the country might grow small quantities of the crops in which other parts of the country specialise. Over the last 100 years Britain has moved from a position where farms in most parts of the country grew a little of everything and kept a variety of livestock – self sufficiency – towards specialisation. It is to be expected that this trend will continue. You are unlikely to find two farms alike but one can recognise broad farming *systems* where farms are similar although varying in detail. These are illustrated in fig. 9.16.

1 There is the division between the arable systems of the south and east of the country and the pastoral systems of the north and west.
2 The arable area can be subdivided into the cereal growing areas and the 'food' producing market gardening area fanning out from a focus around the Fens and the Wash.
3 The pastoral systems can be subdivided into hill farming (sheep and stock rearing) and milk producing areas.

The Common Agricultural Policy of the European Community affects the price of most farm products mentioned in this section. Because the interests of member countries are so diverse, compromise political decisions are reached. They guarantee the prices for various products and this can result in surpluses such as 'butter mountains'. It tends to keep small and inefficient farmers on the land by protecting them from competition within and from outside the E.E.C. So the prices paid to farmers and costs in the shops are not the result of supply and demand alone.

9.16 Agricultural crops in England and Wales

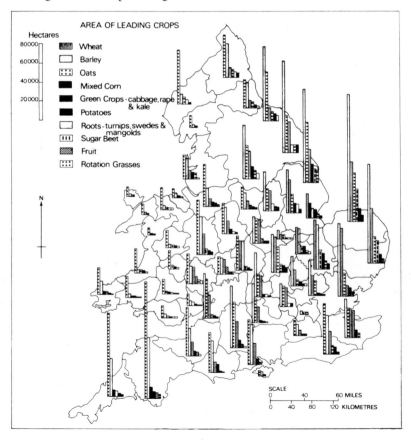

World farming systems

The trends which have occurred in Britain have been repeated throughout the world. They have resulted in a number of farming systems with distinctive characteristics.

1 Plantation crops

The expansion of European settlement and trade led to the production of tropical or subtropical crops for export to Europe. While demand was small it could be met by gathering the product from trees distributed randomly throughout the forests and often miles apart. However, once demand increased it was inevitable that *large scale commercial enterprises* should develop to produce the '*staple*' products required by the world market. Invariably the plantations have been developed with capital and know-how from the developed nations which wanted the products conveniently and cheaply.

Plantations are almost always restricted to the production of *one* crop such as rubber, tea, coffee, bananas, cocoa, sugar, cotton, sisal and other products. This is known as *monoculture*. They usually cover huge areas, for example there is a rubber plantation in Liberia as large as the Isle of Man or Singapore. To all intents and purposes the entire crop is grown specifically for export.

Despite huge investment in machinery, about three quarters of the running costs of plantations go on paying wages or salaries, housing or providing services such as schools or hospitals for the workers. When the number of workers is high and labour charges make up a large percentage of total costs, the concern is said to be *labour intensive*. For example, the Liberian rubber plantation belonging to the Firestone Company of America employs over 21 000 people of whom 12 500 are tappers. There are nearly 13 000 houses, 21 schools, 4 churches, and a hospital serving 60 000 people. Nearly all plantations are labour intensive. A concern which uses comparatively large amounts of machinery and very little labour, such as the cereal growing enterprises on the prairies, are said to be *capital intensive*. In this case the finance is invested in machinery instead of labour. But certain operations on the plantations, such as rubber tapping, and tea picking, cannot be mechanised and payment is made for the skill of the workers who have been able to *specialise* in this one task. Plantations have the size and resources to seek expert advice and opinion to constantly improve their output and efficiency.

2 Ranching and grazing

The growing industrialised and urban populations of western Europe and the USA also created a demand for meat and wool which was met in areas of old European settlement in both the Developed and the Developing World: western USA, interior Australia, South Island of New Zealand, the Karoo in South Africa as well as the llanos of Venezuela, the sertao of Brazil and the pampas of Argentina.

Ranching is a *large scale grazing enterprise* just as a plantation is a large scale crop enterprise. Indeed, ranches are often larger because they are usually situated in less favourable climatic conditions requiring large areas of pasture to support and provide food for the animals. Cattle ranches are rarely less than 1000 hectares in size and those in excess of 4000 hectares are not unique.

About a century ago, when transportation difficulties confined the marketable products to hides, tallow and wool, mixed stocking was common. The introduction of refrigerated ships in 1882 opened up a world market for these products and made it possible to 'specialise' in lamb and mutton production in addition to wool. It was not until the discovery of 'chilling' in 1933 that beef entered long distance trade. Following these inventions it became possible for ranches to concentrate on one animal for one product in the same way as plantations concentrated on one crop. The similarity goes further. The workers, although fewer, specialise on one task, while finance is available for improvements to the pasture, breeding and so on. When comparatively little labour is employed compared to the size of the enterprise both in area and the number of animals, such a farming system is said to be *extensive*.

3 Large-scale grain production

The major staple crops for the majority of the world's population are wheat and rice. They can be compared by testing the hypothesis 'that they represent opposites in farming systems'. Read this description.

On the one hand, wheat is found mainly in countries with a temperate or sub arid climate such as the prairies of Canada, the pampas of Argentina, northern Europe, interior Australia, parts of the USA and Russia, while rice is found predominantly in South East Asia. Only one crop of wheat is harvested each year but two or three crops of rice can often be obtained as climatic conditions are more suitable for growing and ripening. The regions where wheat is grown are sparsely populated, whereas the rice growing regions are densely populated. Consequently most of the rice is eaten where it is produced, but wheat enters widely into world trade. Another consequence of population pressure is the availability of labour and the use of people rather than machines. In wheat growing regions the cost of labour is very high and this, allied to the huge size of the fields and farms which cover hundreds of hectares, makes large investments in machines imperative. Because of the relatively few people involved, the yield of wheat per person is high, whereas in rice growing areas there are so many people cultivating small plots of land that the yield of rice per person is low. On the other hand the yield of rice per hectare is higher than the yield of wheat per hectare simply because of the number of crops that can be obtained from the same piece of land.

Now consider whether the hypothesis is correct – do wheat and rice represent extremes in farming systems?

Aid

The relationships between the Developed and the Developing Countries in raw materials, agricultural and industrial products goes further. The Developed countries commit a relatively small proportion of their budget to 'aiding' less Developed nations. Often the amount and the type of aid is criticised because it is 'tied'. For example, aid could be given to purchase machinery made in the country providing the aid. Whatever the rights or wrongs, aid makes a substantial contribution to the development process. This selection of recent projects financed or supported by the European Community Development Fund (EDF) gives an idea of the scope and purpose of the assistance.

Burundi	Extension of tea growing	Establishment over 7 years of a new tea plantation of 1650 ha.
Cameroon	Establishment of an agro-industrial complex of palm plantation	6000 ha. of selected palms, construction of an oil mill and associated works
Dahomey	Extension of super-structure in the port of Cotonou	Construction of cold storage and refrigeration plants for products of the fishing industry
Gabon	New hotel	Construction of an inter-national standard hotel in Libreville to stimulate tourism
Madagascar	Development of 140 new water points	Construction of 80 rainwater storage tanks and sinking of 60 wells will provide protection for 20 000 people against drought
Mali	Setting up a breeding centre	To develop disease resistant cattle and set up a supervisory network on animal health and nutrition
Senegal	Ship repair yard at Dakar	Creating a ship repair yard for oceangoing vessels (tankers) will provide 3500 new jobs in an area of serious unemployment
Togo	Improvement and asphalting of inter-State trunk route	Encourage increased trade between Togo and Upper Volta, open up backward areas and develop traffic in the port of Lome
Rwanda	Construction of high power electricity line (110KV)	Enables electricity infra-structure of whole province to be grouped into a single system. Furnishes power for tea factory
Mauritania	School building programme	Construction of new class-rooms for elementary education and four secondary schools in regions remote from the capital. Payment of salaries of a number of teachers in first instance
Ivory Coast	Hospital programme	Construction of a 430 bed regional hospital at Korhog, plus a training unit

1 Write a brief essay explaining why these sorts of projects have been selected for the investment of aid. In your opinion, which would make the most effective contribution to development?

Limits to environmental control

This study has indicated the complexity of the factors which have the environment under siege. Nothing can be considered in isolation, as one decision or action sets off a chain of events which have limitless repercussions to the farthest reaches. There are successes and failures, gleams of light and deepest darkness but above all there has been an awakening and growing awareness of the conse-quences of an uncontrolled population growth and an indiscriminate and uncritical exploitation of our planet. Conflict of interests will continue but a third impartial factor will always interpose between the competing factions – the well-being of the environment. May it enjoy good health and long life.

Index